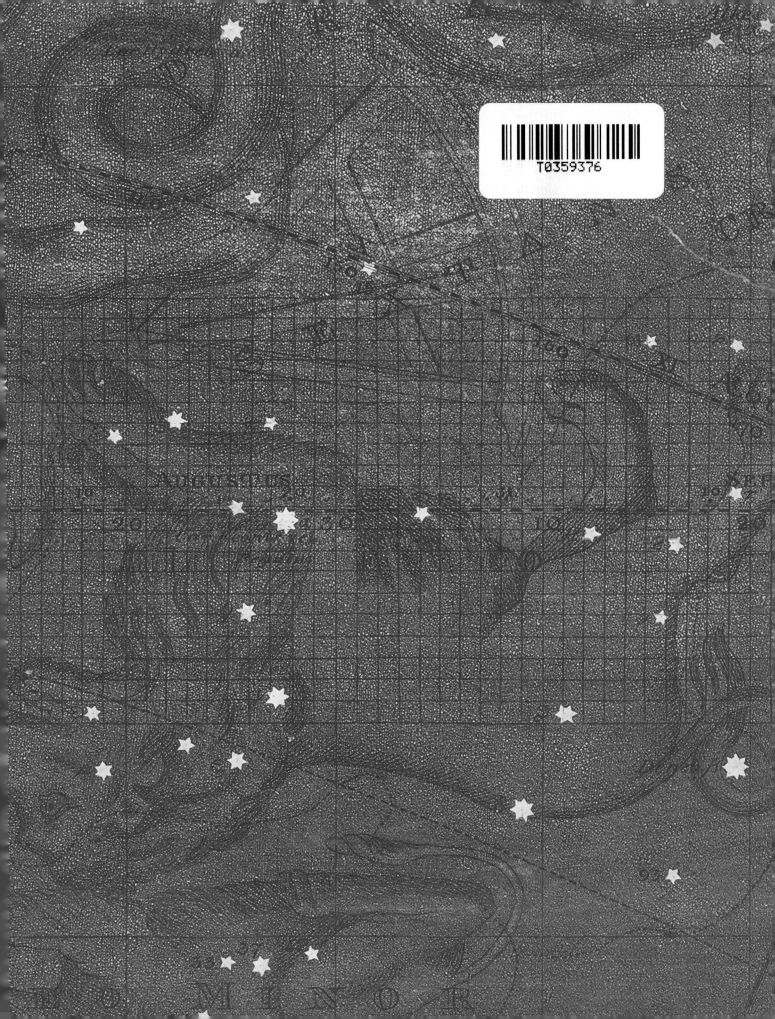

The Long Enlightenment

AUSTRALIAN SCIENCE FROM ITS BEGINNING TO THE MID-20TH CENTURY

Robert Clancy

HALSTEAD PRESS

CANBERRA MMXXI

Published by Halstead Press
Gorman House, Ainslie Avenue
Braddon, Australian Capital Territory, 2612

and

Unit 66, 89 Jones Street
Ultimo, New South Wales, 2007

Designer: Kylie Maxwell, ePrintDesign. Cover concept: Natasha Powers. Printed in China.

 A catalogue record for this
book is available from the
NATIONAL
LIBRARY National Library of Australia
OF AUSTRALIA

The paper for the pages of this book is from a mill certified to have used material from
responsible sources. No old growth timber has been used in its manufacture.

ISBN: 978 1925043 53 2

Illustrations

Except as otherwise noted, illustrations in this book are from the author's collection.
The author and publisher are grateful to the Reserve Bank of Australia for the
illustrations on pages 59, 145 and 154; and to Australia Post for the illustrations
on page 98.

Opposite: Platypuses (*Ornithorhynchus anatinus*) observed on Baudin's expedition,
an engraving from C.-A. Lesueur's picture. See page 24.

Dedication

A SIGNIFICANT OBSERVATION made in this book is of the importance of mentors and role models in the lives of young scientists. Edgeworth David, Charles Martin, David Rivett and Rutherford Robertson are a few who come to mind, each establishing a line of successors. Three such influences shaped my own career. First Ian Mackay, Director of the Clinical Research Unit of the Walter and Eliza Hall Institute in Melbourne. Ian taught me the humanity and science of medicine. My PhD, a consequence of working with him in Melbourne, was in autoimmune disease. Second, John Bienenstock at McMaster University in Canada, who opened the door to the excitement of mucosal immunology, which would become the main platform of my research career. Third, Michael Alpers, Director of the Papua New Guinea Institute of Medical Research. His focus on the relationship between infection and the host response, and on the importance of population studies, opened new dimensions in the way I saw disease.

This book is dedicated to those men, whose vision, integrity and energy inspired and supported me as I sought a way forward, and who became lifelong friends.

—Robert Clancy

The author and publisher are glad to acknowledge

THE SYDNEY MECHANICS' SCHOOL OF ARTS

for its generous contribution to publication, and to record appreciation of

THE ROYAL SOCIETIES OF AUSTRALIA

for their participation, which has enhanced this book.

WARNING: *This book contains pictures of persons who are no longer alive.*

CONTENTS

Does it Matter?

The question has to be asked.
Why write a book on science in Colonial and early 20th century Australia?

Many historians believe they have this covered. Some offer a dismissive observation that apart from the odd ephemeral discovery, science in colonial Australia was all about collecting trophies of a unique natural history for British scientists. From this position their views range through to the limiting idea that any discovery was "utilitarian and localised," as Ian Inkster and Jan Todd put it (in *Australian Science in the Making*, 1988), with a

Planisphaerium Ptolemaicum in *Harmonia Macrocosmica*, A. Cellarius (1660/1708)

Western science as we know it can be given a start date of 1543, the publication date of the idea of a heliocentric solar system by Nicolaus Copernicus, based on observation and limited measurements. Yet a century later in the most decorative of celestial atlases, compiled by Andreas Cellarius, the Copernicus model (over the page) is presented alongside the 1500 year old Ptolemaic view of a geocentric planetary system, an ambivalence reflecting church resistance.

"pragmatic empirical emphasis". American historian Donald Fleming seeing colonial science in the context of the "absentee landlord" reached the unflattering view that natural history for the colonial was "the ideal refuge from the more perilous enterprise of embarking upon theoretical constructs by which he would be pitched into naked competition with the best scholars."

Modern scientists absorbed by their trade see contemporary Australian science as a post-World War II phenomenon. I once shared these views, without stopping to ask "Why?". The answer to my question comes out of concerns I had, to do with contemporary Australian science. After fifty years as a biomedical scientist I found a degree of uncertainty had crept in about the contemporary populism that Australian scientists play above their weight.

Changes I had seen in research and scientific achievement over fifty years were without doubt, dramatic.

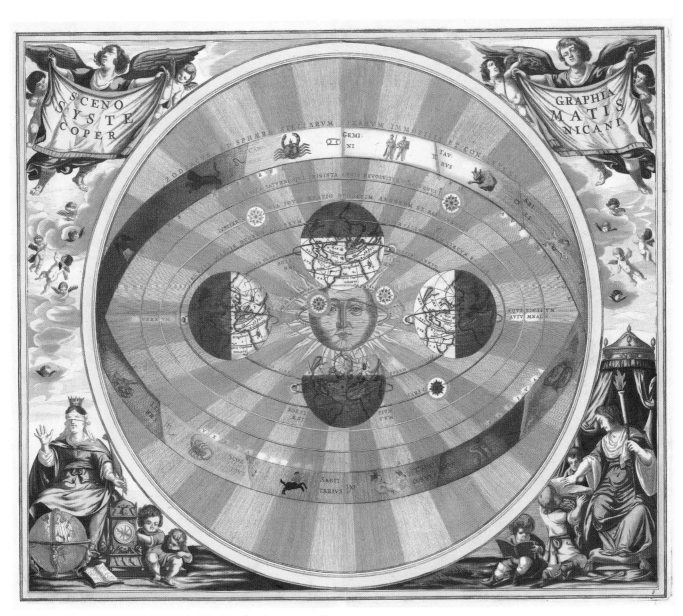

Planisphaerium Copernicanum in *Harmonia Macrocosmia,*
A. Cellarius (1660)

Many were good! Technology and communication in science were revolutionised. My PhD students could not believe that in the 1960s, working on message expression in a metazoan cell system, I was the "automation" in fraction collection – sleeping in the laboratory with an alarm clock to collect the next fraction eluted from a column – or that when I needed a filter to detect immunofluorescence, I made a perspex box (with the help of the Pathology Museum) which I filled with a copper sulphate solution. "P" values were calculated with a slide rule.

Perhaps this just shows my age and is a little self serving, but my point is to contrast the position fifty years later, where the modern scientist can do a sequential base analysis of your genome overnight (while she or he is home with the kids) for a couple of hundred dollars, and the astrophysicist anticipates a data load from the One Square Kilometre Array in Western Australia of 1.5 petabytes, more than the entire Internet! There are so many positives for working in science today, even while the Covid-19 pandemic is consistently reconfiguring our work patterns. My concern was that despite these huge advantages (and perhaps, in part because of them) I had become uneasy with respect to significant discovery, confused over blurred career pathways open to bright, young people and frankly a little angry about ineptitude in translating discovery into product.

Bureaucracy was stifling innovation and research momentum. Considerable and frankly generous funding for research supported process rather than ideas. An industry had arisen around control of research agendas, packed with individuals who had not succeeded in science or who had little understanding of what research is all about. Such an environment was set up for political interference. Witness the stifling of debate on critical issues by the very institutions created to nurture ideas, and the decline of the once great institution of the CSIRO. Translational science was being subjected to short term investment decisions geared to immediate profit – with predictable outcomes for many good ideas. Those same people who had not proven good scientists became involved in running science programmes and would pop up as "advisers" to industry and government. Politically influenced agendas were the consequence.

Young scientists become disillusioned with lack of independence and career paths, finding late in the piece that the narrowness of their research training limits opportunity. Of particular concern was the observation that at every level from performance in science by school students to excellence in research achievement, Australian standards were falling! For example, most Australians awarded the Nobel Prize in science over the last century did the research in the first half of that period. I find that many of my colleagues share these concerns.

How did this situation come about, and what can be done about it? I wondered if there were systemic issues and began looking at how science developed within Australia from Colonial times. I am not a trained historian, though I am deeply interested in the concept that history informs us with respect to the evolution of ideas and activities. Expecting little, I was surprised to discover in our past an overarching philosophical framework that merged science with nation building, and a continuum whereby science and the scientific method influenced most aspects of Australian life from colonisation in 1788 to present times. Recognition of this integration led to the idea that "science in Australia" historically differed from science in Europe. What happened in Australia could never have happened in Europe, and recognising how this came about became a purpose of this book. It was about a philosophy and the men and women who were able to live this philosophy in a remote and isolated continent antipodal to Europe with its hierarchical and restricted structures and traditions. The philosophical framework was delivered by the Enlightenment and colonial Australia was its proof-of-concept experiment. The Enlightenment was a European intellectual movement of the late 17th and 18th century, emphasising reason and individualism rather than tradition. The relationship between Enlightenment principles and the intertwined social and scientific strands of Australian society was akin to the relationship of Newton's laws of motion and gravity to physics, and Darwin's natural selection to biology, as great unifying principles. I argue that the Enlightenment provides a mirror in which the development of science in Australia can be viewed.

"Transit of Venus Observations",
J. Cook (p. 141) in *An account of the Voyages . . . for making Discoveries in the Southern Hemisphere* (vol 2, 1773) J. Hawkesworth.

This page from James Cook's printed journal illustrates the rigidity and selectivity of the sharing of Enlightenment principles within contemporary European society. At exactly the same time as the "officers as scientists" were measuring the transit of Venus, the ordinary folk — the sailors — were stealing nails in order to bargain for sexual favours of Tahitian women.

ROUND THE WORLD. 141

1769. June.

contacts much more than might have been expected. According to Mr. Green,

	Hours.	Min.	Sec.
The first external contact, or first appearance of Venus on the Sun, was	9	25	42
The first internal contact, or total immersion, was	9	44	4
The second internal contact, or beginning of the emersion,	3	14	8
The second external contact, or total emersion	3	32	10

(Morning) (Afternoon)

The latitude of the observatory was found to be 17° 29′ 15″; and the longitude 149° 32′ 30″ W. of Greenwich. A more particular account will appear by the tables, for which the reader is referred to the Transactions of the Royal Society, vol. lxi. part 2. page. 397 & seq. where they are illustrated by a cut.

But if we had reason to congratulate ourselves upon the success of our observation, we had scarce less cause to regret the diligence with which that time had been improved by some of our people to another purpose. While the attention of the officers was engrossed by the Transit of Venus, some of the ship's company broke into one of the store-rooms, and stole a quantity of spike nails, amounting to no less than one hundred weight: this was a matter of public and serious concern; for these nails, if circulated by the people among the Indians, would do us irreparable injury, by reducing the value of iron, our staple commodity. One of the thieves was detected, but only seven nails were found in his custody. He was punished with two dozen lashes, but would impeach none of his accomplices.

CHAP.

Returning to the disparaging opinions generally held on Colonial and post-Federation science, discussed above, I suggest that they fail to recognise two essential truths.

The first truth is that in colonial Australia there was a growth not seen elsewhere – from a gaol and isolated outpost of imperial Britain with one thousand souls, to an independent nation claiming the highest standard of living in the world and a population of 3.5 million, in little more than a century. This did not just happen. The second truth is that there is a continuity in the pattern of science from the time of Cook and Banks up until today, and the Enlightenment is the connection.

To grasp the extent of the contribution of science in the development of the Australian nation, it is instructive to examine science in the context of its economic history. Bernard Attard neatly summarises the economic history of Australia in four phases: the Bridgehead Economy (1788–1820); the Colonial Economy (1820–1930) – a period of explosive agrarian growth; the Manufacturing and Protected Economy (1891–1973); and Liberalisation and Structural Change (1973 to the present). There is a remarkable parallel with changing patterns of science. The first two economic phases align with the Banks era (1770–1820), and the era of applied science (1820–1920).

Economists attribute the "explosive growth" to a "favourable climate" and the inability of Australia's original population to restrain colonial expansion. By

Federation, Australia had by a number of criteria the highest material standard of living in the world. Unlike some other settler societies such as those of South America and Africa, Australia kicked on to maintain that status. Today it is placed only behind Norway and Sweden on the "Human Development Index". (Norway's and Sweden's positions are due to clever use of science.)

Attard explains Australia's extraordinary rise and retention of living standards in economic terms, as its capacity "to respond to successive challenges to growth by finding new opportunities for wealth creation with a minimum of political disturbance, social conflict or economic instability, while sharing rising national incomes as widely as possible". He concludes that the Australians' capacity to change is the basic ingredient of their prosperity and hope for the future.

It is a fundamental concept of my thesis, that the growth, success and capacity to adapt identified by economists are an expression of Enlightenment principles imbued from the time of Banks and Cook, made possible by exceptional individuals who came to make their lives in a remote and harsh land, and of the scientific processes they used to survive and profit.

Back to the "truths". Growth depended, as did survival, on remarkable people, solving a constant stream of novel problems, using the scientific method. They first identified a problem that had to be solved, then developed a strategy on how to solve it by empirical data and experiment – the classic scientific method. Here science was not owned by an elite scientist but rather was a process involving "ordinary" people. We "professional scientists" can be quite precious about what science is, so it is useful to look briefly at what Western science is all about.

The concept of science is a simple one. It is antithetical to mysticism and tradition. It involves analysing the natural world by observation and experiment. The way science worked in Europe was very different to what happened in 19th century Australia. Populations in Europe had adapted to conditions over long ages. There was a hierarchy based on inheritance and layered social strata. There was little opportunity to stand out and make a difference. Innovation was for the privileged

few. Britain, developing domestically within European limitations, would continue to benefit in science from Australians and British born scientists who spent time in Australia, whose success flowed from Australian opportunity.

The association and link between science in Australia and in Britain is clear – including the pattern of influence of the Enlightenment, and (at least in the early Colonial period) the fit due to its geography, isolation and unique biology, that conformed Australia with the strengths of British science. Review of the history of Western science with a focus on Britain's part shows that such relationships were not happenstance. Importantly, understanding what science meant and how it developed to become an idea of the people and part of everyday life helps us appreciate how science became part of the growth of Australia.

Modern Western science began in northern Italy in the shadow of the Renaissance and evolved over 250 years with empiricism at its centre, trumping mysticism in relation to Man, Earth and the Cosmos. Fra Mauro collected observations of geography in 1450 to challenge the Ptolemaic view with his extraordinary map. In 1543 Andreas Vesalius' observations of anatomy based on human dissection replaced Galen's texts based on apes, while Nicolaus Copernicus' heliocentric conception of the planetary system challenged not only Ptolemy, but the dogma of the Catholic Church. These pillars of empiricism were followed by Francis Bacon who developed and promoted experiment, inductive thinking and the scientific method (summarised in his *Novuum Organum,*1620); and by two foundation experimental scientists, William Gilbert, author of *De Magnete* (1600), and Galileo with his wide ranging discoveries based on observation and experiment which yielded new laws of motion and physics especially as they related to astronomy (1633). Galileo's experiments paved the way for Newton's codification of classical mechanics. Newton's laws of motion and gravity established for the physical world and the cosmos a unifying principle expressed in mathematical formula and published in *Philosophiae Naturalis Principia Mathematica* in 1687. Melding with an emerging philosophy derived from the Reformation, that promoted the importance of the indivi-

"Terre de Diemen. Navigation" in *Voyage de Decouvertes aux Terra Australis*, Lesueur et Petit (1824)

The influence of science and the Enlightenment is powerfully reminded by the illustrations in the published journals of the great French expeditions into the region of Terra Australis. Thus the voyage of Nicolas Baudin (1800–03) included fourteen naturalists with expertise in natural history including anthropology and ethnology. It is not surprising that they had a scientific interest in the lifestyle characteristics of Australian Aborigines.

dual, and a liberalism championed by philosophers such as John Locke, they led to the Enlightenment movement.

This intellectual and philosophical movement developed through the 18th century, promoting the idea that humanity can be improved by rational thought and change. In the highly regulated and hierarchical European society of privilege, the tensions created by the Enlightenment were major factors in promoting revolution and explosive change. Colonial Australia was very different.

British prominence in the promotion and exportation of Enlightenment principles was based on solid grounds, due to Britain's leadership at the interface of science and philosophy, and its imperial strength. Bacon had established a framework for the modern scientific method, and William Gilbert was his first disciple. In the 17th century, Newton was joined by a crowd of scientists who forged ahead in the natural and physical sciences; to name a few, Robert Hooke, Robert Boyle, William Oughtred, James Gregory, Edmond Halley, John Ray and William Harvey.

Bacon had recognised a need for an independent specialist society – and one was created by a group of English scientists that included Robert Boyle and Christopher Wren. The Royal Society with its motto "Base nothing on words" began in 1660, with great scientists such as Robert Hooke and Isaac Newton becoming fellows. Its *Philosophical Transactions*, begun in 1665, became the leading publisher of scientific discovery for the next 200 years, including contributions from Newton and Darwin, until *Nature* emerged in 1869 from the profusion of scientific journals that appeared when publishing restrictions were lifted in the mid-19th century.

One of the first activities of the newly formed Royal Society was to publish "Directions for Sea-men Bound for Far Voyages" (Lawrence Rooke, 1661), which made clear expectations of sea captains to make scientific observations such as "declination of compass or its variation – frequently making withal the latitude and longitude of the place – and setting down the method." Developing on the back of 17th century trade companies which dealt in cargoes across the world, Britain's imperial interests shaped science and produced discipline strengths. These influences determined patterns of science that unfolded in colonial Australia – and indeed, continue into today's world. Britain's powerful control of international sea routes would consolidate following success in the Seven Years' War, fought across the world in the 1750s and 60s. Astronomical information to aid accurate navigation, cartography and time keeping were needed especially with respect to the calculation of longitude. The Royal Society was closely aligned with both navigation and natural history, areas of science that became the face of enlightenment for imperial Britain in the 17th and 18th centuries. Indeed, the Society sponsored expeditions with scientific agenda including James Cook's first voyage into the Pacific in 1768 to observe the transit of Venus. Joseph Banks – naturalist aboard Cook's *Endeavour* and catalyst of research into the unique flora found in Australia – would become President of the Royal Society for over forty years.

It is not surprising that the Royal Society supported the establishment of the Royal Observatory at Greenwich in 1675 to 'rectify the tables of the motion of the heavens and the place of the fixed stars so as to find out the so much deserted longitude". The President of the Royal Society was a member of the Board of Management of the Royal Observatory. The Observatory's first Director, John Flamsteed, managed to construct a star catalogue containing 2,935 stars with unprecedented accuracy. However, it took nearly a hundred years for

an accurate set of tables to be produced to enable the angle between the Moon and a fixed star to be correlated with Greenwich time – a method used by James Cook in 1768 and many navigators, until 1850, when accurate and affordable chronometers became available.

Hans Sloane, a leading physician and amateur naturalist, was made a Fellow of the Royal Society in 1687. He provided land for the Chelsea Physic Garden in 1673, as a teaching laboratory for apothecary apprentices. It would quickly become a "botanic garden" with the world's greatest collection of exotic plants from the new world. It became famous for its contribution to British control of trade by acting in a shuttle fashion for redirection of exotic commercial crops, such as cotton to Georgia, and for its role in international seed exchange programmes. Its connection with the Royal Society was maintained through the annual supply of fifty herbarium samples and a close working involvement with Joseph Banks, who was also advisor to the Kew Botanic Gardens from their inception in 1772.

New era colonial Australia brought together a near perfect storm of factors to test Enlightenment principles. Important in the mix was the nature of immigrants – who all had points to prove. The list includes minor criminals, many sentenced for political crimes, young men and families of variable wealth with a limited future in Britain, and professionals attracted by health and opportunity. The underpinning desire was to prove themselves in a remote and harsh land with unique and unexpected challenges, but with perceived opportunity to prosper and survive. The experiment was a perfect test of the German philosopher Immanuel Kant's view of the Enlightenment as "freedom to use your own intelligence". It was about being prepared to have a go, to create a life not otherwise available. For some such as botanist Allan Cunningham, Governor Thomas Brisbane and Secretary Alexander Macleay, it was an unusual opportunity to become involved in the new science of "the Antipodes". The common denominator for all who came was the unique environment of Terra Australis. It was "being there" that enabled remarkable individuals to seize opportunity, exploit that using principals of science to overcome challenges, and shape Australian society.

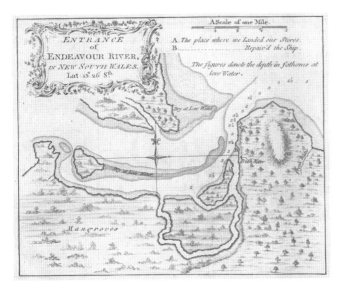 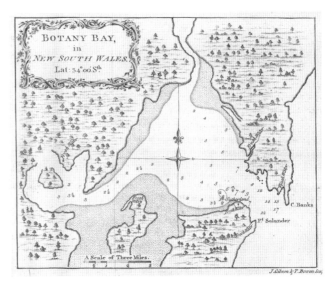

"Entrance of Endeavour River", "Botany Bay in New South Wales" in *An account of the Voyages . . . for making Discoveries in the Southern Hemisphere* (Vol 3 1773) J. Hawkesworth

The *Endeavour* spent ten days at Botany Bay in 1770, during which Western science was aligned with the challenge of Terra Australis. These maps and the picture highlight the talents of James Cook in "navigational science" and Joseph Banks in "natural history". Cook was more than a master mariner; he was a scientific cartographer with skills and application in the range of physical sciences required to navigate and record discoveries in the Southern Hemisphere. These sketch maps record the two locations along the east coast where the *Endeavour* spent time. The map of Botany Bay includes the entrance of George's River, with soundings and latitude. The Grevillea species pictured was grown in England from seed found by Banks at Botany Bay. These early Australian botanical specimens stimulated enormous interest and were the major reason for the public success of serialised botanical magazines. The most important was *The Botanical Magazine* by William Curtis, began in 1787 (which continues today as the official Kew publication). Over the next forty years, Australian plants constituted about 25% of species illustrated.

Before colonisation this environment had already nurtured scientific progress by Aborigines. Unlike the new colonists they had to solve problems without access to imported products and ideas. As scientists do, they relied on observation and experiment to develop detailed knowledge of nature, as well as distinctive technologies which addressed challenges of survival – such as woomeras, fish traps and land management practices. R.W. Home captures this concept: "Aborigines' systems of classification gave them a remarkable intellectual mastery of their surrounds."

After the colonists took it over, an unexpected measure of this environment was the impact on Britain of returning British "Australians" and native Australians who went there. Frederick Wolseley and Herbert Austen whose innovation in Australia led to the invention of mechanical shears, returned to Birmingham to pioneer the British motor car industry. Howard Florey, born in Adelaide, took a Rhodes scholarship in Oxford because appropriate science training was not then available in Australia, and developed penicillin for clinical use – a world changing event.

As they did in Florey's time, and before European colonisation, new challenges will continue to demand scientific responses. Even as this book was nearing completion, Australia endured the severest bushfires on record, and a deadly new strain of corona virus tested authorities and scientists in ways we will examine later.

Today, we know the names of a few whose achievements in science contributed to the development of a nation, but only if they appear on stamps or banknotes. We once learnt about others at school! What is not understood is the broad role science played in achieving the growth essential for nationhood. There was a buzz in the community about science and discovery. Individuals came together to solve problems in rural science (such as entrants in the competition that led to Ridley inventing a stripper to harvest wheat in quantities too

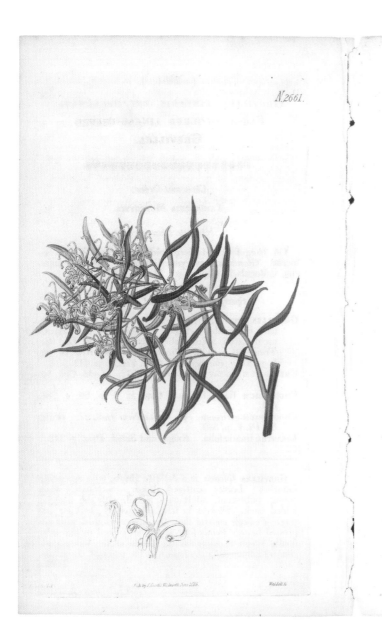

(2661)

GREVILLEA LINEARIS. *var.* INCARNATA.
FLESH-COLOURED LINEAR-LEAVED
GREVILLEA.

****

Class and Order.

TETRANDRIA MONOGYNIA.

Generic Character.

Cor. irregularis. *Antheræ* apicibus concavis, corolla im-
mersæ. *Glandula* hypogyna, dimidiata. *Folliculus* supe-
rus, unilocularis, dispermus. R. BR.

Specific Character and Synonyms.

GREVILLEA *linearis ;* foliis lineari-lanceolatis acutis mucro-
natis marginibus refractis, racemis abbreviatis erecti-
usculis, stylis apice glaberrimis. *Brown in Lin. Soc.
Trans.* 10. *p.* 170. *Prodr.* 376. *Hort. Kew. ed. alt.*
l. *p.* 205. *Roem. et Sch. Syst. Veg.* 3. *p.* 411.
EMBOTHRIUM lineare. *Bot. Repos.* 272. *Lodd. Cab.* 50.
858.
EMBOTHRIUM linearefolium. *Cav. Ic.* 4. *p.* 59. *t.* 386.
f. 1.
EMBOTHRIUM sericeum. *γ. Smith New Holl.* 27. *Willd.
Sp. Pl.* 1. *p.* 539.
LYSANTHE linariæfolia. *Knight and Salisb. Prot. p.* 118.

GREVILLEA *linearis* is a delicate shrub, with spreading
branches. *Leaves* scattered, linear, mucronate, when
young, pubescent, adult ones naked, with the margins
folded back. *Flowers* in terminal racemes, looking one
way. *Pedicels* shorter than the flower, clothed with ad-
pressed hairs. *Petals* four, equal, all turned to one side,
finally revolute, hairy on the inside at the lower part.
Anthers immersed in the hollow at the point of the petals,
before

"Grevillea linearis" (1770) in Curtis's *Botanical Magazine* (2661)

great for manual labour). Much of the early teaching in basic sciences such as chemistry and mathematics was through these organisations begun by members of the public, beginning with the Sydney Mechanics School of Arts in 1833.

The value and role of mechanics institutes in Australian society has been misunderstood by some as "providing an infrastructure within which the philosophical and 'royal' societies could grow" (Inkster and Todd, *Australian Science in the Making*). Views that failed to recognise the unique character and universality of science in colonial Australia. In New South Wales about 18,000 attended 269 Mechanics Institutes in 1897. The Sydney Mechanics School of Arts had a membership of 5,000 in 1891, with significant government subsidies, with lectures and courses across a wide scientific spectrum, but a focus on the practical and technical. These institutions morphed into the full spectrum of technical colleges and universities that appeared in the late 19th and 20th centuries, beginning with the successful Sydney Technical College (precursor to the University of NSW).

Variations on the theme occurred across the colonies. For example in South Australia between 1856 and 1884 the scientific community including the Adelaide Mechanics Institute and the Royal Society of South

Australia was organised within a government-financed South Australia Institute, which functioned as a centralised scientific culture aiming at benefits of scale.

The first scientific society in Australia began in 1821 as the Philosophical Society of Australasia with Governor Brisbane as its President, "to enquire into various branches of physical science of this vast continent and its adjacent regions". It had a precarious early history not unrelated to the politics of the time, but from 1850 its successor took a leadership role in coordinating science, becoming the Royal Society of New South Wales in 1866. The peer reviewed *Journal and Proceedings of the Royal Society of New South Wales* includes a remarkable chronology of scientific achievement across colonial Australia (and well into the 20th century, until specialist societies became established). Similar societies were established in Hobart (1843) and Victoria (1854), South Australia (1853), Western Australia (1914) and Queensland (1884). The journals' indices reflect the breadth of scientific interest amongst amateurs of science and early academics in the era before specialisation.

The growth of scientific interest across the colonies can be seen in membership of the colonial Royal societies. In 1990 the New South Wales society had 374 members; Victoria's 136; South Australia's 74; Queensland's 100; and Tasmania's 111. The high proportion of members without a formal science background reflected the broad community interest in science. Before 1890 only a minority of members of the Royal Society of New South Wales had professional scientific status.

More specialised societies focussing on particular areas of science appeared in the latter part of the 19th century, for example the Linnean Society of New South Wales founded in 1874 to study natural history and the Astronomical Society of South Australia begun in 1891.

Colonial governments sponsored meetings to address particular science challenges such as control of rust disease in wheat and babesiosis in cattle in the 1890s. Public science was further promoted by professional overseas recruits to government and university positions. For example, Ferdinand von Mueller, Director of the Melbourne Botanic Gardens, reported 200,000 visits to the Gardens in 1858/59, while Frederick McCoy – Professor of Natural History at Melbourne

Fish Traps (National Museum of Australia)

Above: Simple but effective in design, fish traps vary to suit local conditions.

Boomerang (National Museum of Australia)

Below: There is no better example of Aboriginal science based on observation and experiment than the ubiquitous boomerang. The boomerang's operation is analysed as an example of gyroscopic precision with the airfoil causing it to fly, the higher aerodynamic lift of the top end creating a torque causing the angular momentum to proceed. But that knowledge was not necessary to its invention and improvement, which were based on observation.

University – had 32,000 visits to the Museum he curated in the University grounds. Libraries attached to Mechanics Institutes were popular, while newspapers regularly reported on local scientific discoveries. The Hobart Mechanics Institute with 400 members in the late 1850s loaned about 10,000 books each year. Science was owned by the public!

For fifty years until his death in 1820, Joseph Banks was the dominant influence on all discussions relating to science in colonial Australia, in part through his Presidency of the Royal Society. It was Banks who established natural history as a core discipline that continues today. He recommended Allan Cunningham's appointment to New South Wales in 1816 as a collector for Kew Gardens – a watershed appointment. Cunningham would spend the best part of twenty years in Australia combining geographical discovery with foundation academic contributions to the collection and classification of Australian flora. His career was a turning point, where scientists began calling Australia "home" while still recognising an attachment to Britain. He actively participated in local scientific groups, publishing both geographic and natural history discoveries. This turning point coincided with the end of the period of the Bridgehead Economy of the early settlement years, and the "beginning of growth".

Continuing interaction between Australia and England influenced the scientific development of both. While German scientists developed laboratory science with a revolution in the study of plant structure and function, the excitement of the new botany drove a British focus on phylogeny and classification. Despite its limitations, emphasis on classification – especially by applying the "natural criteria" promoted by Robert Brown (who on Banks' advice had accompanied Matthew Flinders), was a valuable precursor to studies that led Charles Darwin to develop his theory of natural selection. The focus on species and their natural environment would spawn the development of ecology and environmental science – opportunities not lost on Australian scientists aware of new species in unknown environments.

The first scientific expedition to central Australia in 1894, known as the Horn Expedition, involved a remarkable inter-colonial collaboration involving scientists from Adelaide, Melbourne and Sydney. Outcomes of the experiences gained included study of the impact of the environment, and the new science of anthropology. There followed a series of similar expeditions, culminating with the American-Australian Scientific Expedition to Arnhem Land in 1948 led by Charles Mountford. This co-sponsored expedition highlighted Australia's contribution to natural history and ecology squarely in an international light and a post World War II shift in Australia towards an alignment with American science.

The rise of anthropology was an extension of natural history, predicted when Linnaeus included man in his binominal classification of plants and animals. It was brought into focus by Darwin's halting inclusion of man in his ideas of natural selection and became a serious scientific study in Australia with the meeting in Alice Springs of Baldwin Spencer and the Alice Springs telegrapher F.J. Gillen, during the Horn expedition. It launched the zoologist Spencer into committed study of anthropology in central Australia. To an extent this was a beginning of anthropology, a discipline where Australia would provide scientific leadership.

Evidence of continuity of the influence of the Enlightenment on the rigour and pattern of Colonial and early Federation science can be found in review of scientific publications. Before 1900 most scientific publications were in the proceedings of the colonial Royal Societies (and in newspapers), though some studies in astronomy and natural history were reported in proceedings from specialist societies in Europe (and in Australia – the Linnean Society of New South Wales, and the Astronomy group in Adelaide).

Change to academic leadership was dramatic, largely due to the recruitment of Archibald Liversidge as Professor of Chemistry at Sydney University. Liversidge made a difference across the spectrum – he improved the school curriculum in science, fought battles within the University in support of science, and provided a leadership role through his involvement with the Royal Society of New South Wales. Perhaps his major contributions were promoting science as public property and in bringing together the professional scientific community throughout Australia. This he achieved by playing a key role in the creation of the Australasian Association for the Advancement of Science. This powerful group was immensely successful in helping Australians understand that science made Australia great, promoting the view "that almost every aspect of our life is touched by science. Without science our way of life would be almost unimaginable and sustainability of life in the future would be impossible."

After 1920 it is more difficult to monitor patterns in science through publications of the state Royal Societies. University Departments, institutes and the CSIR moved away from generic local publications to international and specialty journals. Notwithstanding, the Royal Society publications continued to reflect the prominence of natural history.

Astronomy had become "national" with the opening

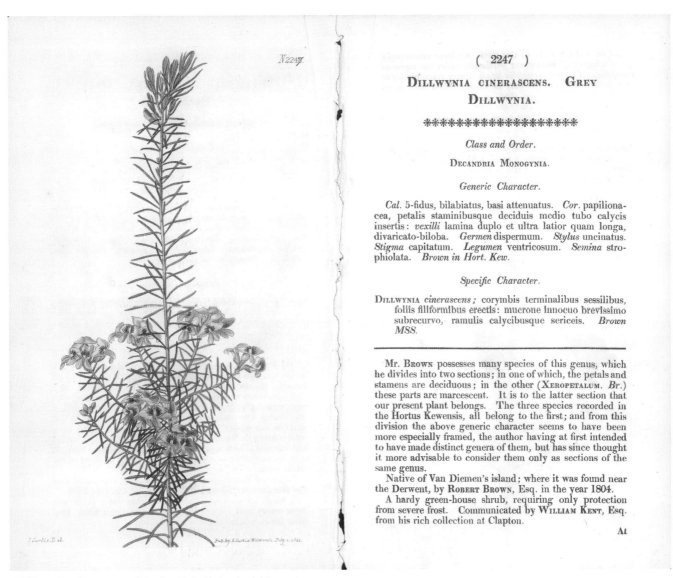

"*Dillwynia cinerascens*" in Curtis's *Botanical Magazine*.

Robert Brown was the dominant botanist in the world through the first half of the 19th century, shaping the pattern of botanical research for Britain and Australia as classification and descriptive botany. His status was due to his studies of Australian flora. He became best recognised as a botanist with Matthew Flinders's circumnavigation (1801–02) and for his studies of the family *proteaceae* (now known to contain 1,700 species, predominantly in the Southern Hemisphere), which he published in his *Prodromus* in 1810. His continuous association with Australian botany extended to 1844 with his addendum to Charles Sturt's published journal which details an expedition into the centre of Australia. The specimens included here were found during his voyage with Flinders and Bauer; *Dillwynia* is a Tasmanian native Brown collected in 1804.

of Mt Stromlo Observatory in Canberra. Optical astronomy was transformed with the new technologies such as spectroscopy, and botany shifted its focus to environmental impact including shorelines. A major area of research involving plants and animals including human beings, in health and disease, was the study of host-parasite relationships.

Recognition of the importance of microbes thanks to Pasteur's germ theory revolutionised biology. Its introduction into Australia by Pasteur's nephew Adrien Loir, who prevented anthrax in Australian sheep by immunisation, and John Ashburton Thompson's demonstration of the practical value of scientific principles of epidemiology in controlling bubonic plague, began a continuous stream of biological inventions that remain dominant in contemporary science.

Geology shifted to dynamic studies of processes such as glaciation and geomorphology, while continuing to address big questions of evolution (Were there metazoan fossils in pre-Cambrian rocks?) and plate tectonics.

Early chemistry began as analysis of natural products but moved to more complex studies of metallurgy and atomic structure. Anthropology moved on from evolutionary anthropology to social anthropology, concentrating on studies of cultural and racial interaction. Physics and astronomy stayed in the international sphere with studies on the ionosphere, photography, radiation and wireless. Advantages of a southern geography were extended with studies in the Australian territories of Antarctica. While there was no dramatic shift away from traditional strengths of natural history and navigational science inherited from Joseph Banks and James Cook, globalisation through communication and an enthusiasm to compete on the world stage meant a blurring of these traditional patterns in the post World War II era.

This book is about heroes! There isn't a place in the coming chapters for every hero who made an important difference. However there is, I trust, a thought provoking array of extraordinary yet largely unknown people who forged a nation by using scientific methods to solve problems. Collectively they established a narrative of applied science influenced by the principles of the Enlightenment and the "freedom to use one's own intelligence". Science was public property and for 150 years Terra Australis was an experiment of the Enlightenment. The idea of this book came from a pondering after a working life in science, as to whether contemporary science is in good health. To understand the question better, I searched the patterns and details of science in Australia and found an unexpected and amazing story of individuals – most previously unknown to me – who used scientific method to shape the Australia of today. In their time science was an integral part of life.

If the misgivings I share with others about science today are warranted, fuller knowledge of an earlier period when results were more conspicuous may be just as useful as criticisms, or more useful. To benefit from history we don't have to wind the clock back; and we don't have to pretend the past was a golden age in order to find guidance from it for our own time. It is a fact that in important ways conditions have improved for science – the enhanced facilities and capacities mentioned on the first page for instance. The opening of science to women, who until current generations were mostly excluded by rules and social norms, is an example of overcoming a handicap which impeded science in the period examined by this book. A society which succeeded brilliantly, notwithstanding handicaps, surely offers much from which we could beneficially learn. Freed of such handicaps, can we not aspire to do better?

The story of Australian science can conveniently begin with the influence imposed by two men – James Cook and Joseph Banks. They mediated the spirit of the Enlightenment, and initiated patterns of scientific endeavour that became bedrocks of Australian science – navigation science and natural history.

In the Beginning

James Cook earnt his opportunity to lead the scientific expedition of the **Endeavour** *through notable contributions to marine survey made possible by a high level of astronomical knowledge.*

His broad command and use of navigational science place him beyond being just the first European to discover the east coast of Australia (which was simply a question of "when?") and stand him as the foundation natural philosopher of Australian science. Cook was not just another navigator. He had sought out new methods such as plane table surveying, which he learnt from the surveyor Samuel Hatland, to combine land based triangulation with offshore running surveys, by which he could compile extraordinarily accurate and valuable charts.

By 1768 when Cook left on the first of his three voyages in the Pacific, he had published a valuable portfolio of charts with sailing directions for two thirds of the coast of Newfoundland. Following Cook's death, Henry Roberts in 1780 produced a summary chart of the Pacific Ocean to accompany the journal of Cook's third Pacific expedition, highlighting discoveries of Cook. This chart included all major island groups in the "right" place. As a small scale map, it is remarkably similar to an accurate modern map of the Pacific.

The voyage of the *Endeavour*, begun in 1768, during which Cook discovered and charted the east coast of Australia, was a scientific expedition promoted by the Royal Society. The primary goal was to record the transit of Venus across the face of the Sun on 3 June 1769 in order to determine the solar parallax, representing the true diameter of the Earth, so the distance of the Earth from the Sun could be calculated with trigonometry, using a formula designed by Edmund Halley. Cook and the expedition astronomer were unable to obtain accurate readings due to the blurring of transit edges, a failure shared by observers of the transit at sites around the world.

Their other astronomical observations included the observation of the transit of Mercury off the east coast of Australia, and the first long voyage verification studies of predicted positions of Sun and Moon recorded in the first *Nautical Almanac* published in 1767. Cook used both lunar distance and Jupiter moons as "celestial clocks", and estimated lunar distance accurate to half a degree. These calculations were used to chart the Australian coast.

Part of the attraction that led to James Cook's appointment was his broad commitment to science. He is remembered for his contribution to solving the riddle of scurvy, known now to be caused by vitamin C deficiency. Scurvy regularly decimated long haul crews and death rates of thirty to fifty per cent were common. Cook's "clinical trial" testing malt, "potable soup", sauerkraut and rob (a boiled down citrus preparation) was confused in its construction. It led him to conclude that malt was the best antiscorbutic.

Such was the respect for Cook's scientific abilities, that this view held sway – indeed the Royal Society rewarded him with its Copley Gold Medal. In fact it was because of his attention to detail – emphasising fresh air, cleanliness, clean cooking pots, and frequent stops where fresh vegetables were collected – that Cook had no deaths from scurvy on his 1768–72 circumnavigation. His "antiscorbutics" contained little or no vitamin C and made little difference. The problem at that time was there was no concept of specificity in the causation of disease and any factor "changing the balance of humours" was considered adequate to account for differences in health and illness.

Cook made a major difference to the health of seamen in a general sense, despite delaying the use of specific antiscorbutics in the British navy. Despite his understandable confusion, his approach to managing

A view of CAPE ESPIRITU SANTO, on SAMAL, one of the Phillipine Islands, in the latitude of 12:40 Nº. Bearing WSW distant 6 leagues. In the position he represented his Majestys Ship the CENTURION engag'd and took the Spanish Galeon call'd NOSTRA SEIGNIORA DE CABADONGA, from ACAPULCO bound to MAN...

"The Centurion engaged and took the Spanish Galleon call'd Nostra Seigniora" in *A Voyage Round the World . . .* by George Anson, R. Walter (1748)

Anson's aggressive foray into the Pacific — disastrous in human terms with more than 80% of a complement of 1,800 lost to scurvy — was a prelude to the "Seven Year War", a conflict that established Britain as the leading imperial power, and heir to rich takings in the Pacific region, with export of the European Enlightenment as "compensation".

the conditions of sailors influenced Gilbert Blane who introduced lemon juice for all navy sailors in 1785. It was an effective anti-scorbutic which gave Britain an edge in the Napoleonic Wars by enabling ships to remain at sea for long periods.

Not only did James Cook bring the navigational sciences to Australia, but his influence persists today as Australians maintain leadership roles in physics and astronomy.

Joseph Banks (1743–1820)

Joseph Banks was a naturalist and patron of Australian science – but he was much more than that. He defined the science agenda in Australia for fifty years, was the gate through which all scientific endeavour passed, and the catalyst for others wanting to progress in Australian society. He saw Terra Australis as a laboratory for English scientists but, with time, those he employed as his technicians in that laboratory saw much more. Robert Brown took the opportunity to become the leading British botanist, while Allan Cunningham understood the potential for an Australian independence.

Banks had all the trappings a wealthy amateur

naturalist needed – private tuition, talented travelling companions, and managerial skills and contacts. He was directly responsible for England taking the leading role in world botany – to the extent that he facilitated the acquisition of Linnaeus' herbarium following Linnaeus' death, and ensured imperial England used its influence to flood Kew Gardens with exotics from the far reaches of empire. Before his life changing trip with James Cook, he had travelled to Newfoundland and in 1766 been elected a Fellow of the Royal Society. He was the Society's President from 1778 until his death in 1820, giving him extraordinary power in the world of science.

After his return from Terra Australis he continued to travel in search of new species – to Wales, islands off Scotland, and Iceland. He held "Antipodean court" in his house in London, sending botanists to Australia to collect, classify and store specimens for transfer to London. Such was his impact on imperial science that over 7,000 new exotics were imported during King George III's reign.

His interest in farming and business extended his involvement to many aspects of the development of colonial New South Wales, particularly as an adviser on agriculture and wool production. Joseph Banks was the glue to world botany and the foundation stone for Australian natural history.

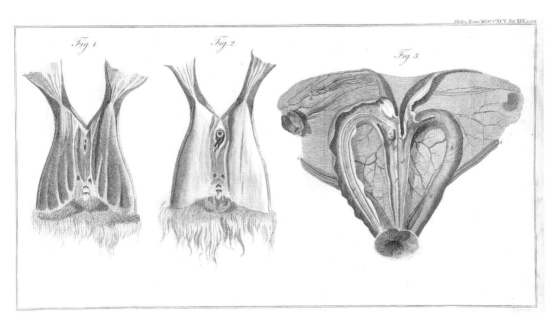

"Some Observations on the Mode of Generation of the Kanguroo" in *Philosophical Transactions* **(of the Royal Society) (March 5, 1795), from E. Home**

The "Kanguroo" became an immediate icon of antipodean natural history. It was presented to the British public by a print included in Cook's published journal — a bestseller that would be reproduced in many languages and reprints. George Stubbs, known for painting horses, painted a stuffed skin of a kangaroo shot when Cook was beached at the Endeavour River — the print was copied from Stubbs's painting. This was a period in France and England when there was intense interest in comparative anatomy as part of a period of pre-Darwinian interest in the idea of evolution. In the Royal Society proceedings, Home adds to this discussion with a description of a unique anatomy of the female "Kanguroo's" reproductive system, relating this to an unusual sequence of fertilisation of the ovum. He includes comments from the field made by a Mr Considon while serving as assistant surgeon in colonial New South Wales.

Natural History:
The Enlightenment Epiphany

The idea of natural history has changed.
In 100AD Pliny's **Natural History** *included everything from astronomy to medicine with even a touch of superstition, in encyclopaedic manner.*

Renaissace men, the "natural philosophers", differentiated empiricism – experience derived from observation – from mysticism or irrational thought.

This was the beginning of modern Western science though the term scientist to describe those who researched was not introduced until 1833. These same Renaissance men added a third area – nature – to the traditional knowledge of their times, which already included the humanities and religion.

Knowledge of nature was either *descriptive* or *analytical*. The latter became known as "natural philosophy" and included physics and astronomy, while descriptive studies of nature included plants, animals and the Earth. Earth studies were late starters, but by 1800 they had their own discipline, geology, an identity encouraged by the Industrial Revolution with its demand for the economic recovery of minerals. A century earlier natural philosophy as it related to the Earth and the Cosmos had a unifying principal in the form of Isaac Newton's laws of motion and gravity. This was not the case for "natural history", which had lingered as materia medica, herbals and physic gardens, little changed from the time of Galen, reflecting the dominance of medicinal value in botany. The pressure for change came with two events that took the public's imagination in Europe in the 16th century: the printing press mass producing text and accurate images and the flood of exotic plants to Europe from new trade routes to the Americas and the East. The outcome was that physic gardens become botanic gardens, and herbals became florals. The added pressure from new collections cried out for a comprehensive classification system so essential for botany to progress in the scientific world.

The breakthrough came when John Ray, an English parson-naturalist, developed a classification system, with *species* as the taxonomic unit. He defined species as "never born from seed of another". Ray changed the world of natural history, basing his classification on physiological and structural characteristics of the plant. At the same time, his system recognised family relationships and distribution characteristics. Ray's classification of 18,000 plants was published between 1686 and 1704 in his epic *History of Plants*.

Carl Linnaeus developed Ray's system into a more comprehensive and practical one, based on morphological characteristics of male and female reproductive organs, in a binomial format. His *Species Plantarum*, published in 1753, included 7,700 species of plants and 4,400 of animals. Including man as an animal was a controversial move at the time. Linnaeus was of importance to Australian botanical research in two main ways. First, it was his system that Joseph Banks would use to classify the collection of plants that he brought back from Australia and the South Seas to England in the *Endeavour* in 1771, creating a scientific boost to world botany. Second, through the intervention of Banks, and following the death of Linnaeus, England obtained the massive reference herbarium used by Linnaeus in establishing his core database. This was such an important event that together with Banks' personal contributions, England took the pivotal place in world botany.

With an apparent never ending flow of exotic plants from the new worlds, taxonomy was the name of the game and the dominant scientific interest. Some have suggested that for much of the 19th century the strength of focus on taxonomy disadvantaged British science in comparison with the ground-breaking studies of German botanists on botanic structure and physiology. The British influence with its focus on taxonomy dominated Australia's scientific botany into the 20th century except that two other powerful strands flourished: economic botany as it related to agriculture, and study of Australian flora, ecology and evolutionary biology with the focus on natural selection and the environment launched by Charles Darwin's *On the Origin of the Species* in 1859.

An important shift in approach to taxonomy was adoption by early botanists working in Australia (notably Robert Brown) of an improved system of classification based on natural characteristics published in 1789 by French physician-botanist Antoine de Jussieu. More scientific in its attention to physiological characteristics, this system was immensely important to the study of inter species relationships and evolutionary biology. The human species was in the mix and botany was finally freed from restrictions imposed in the materia medica era. The playing field was cleared for Charles Darwin, and for the impact his ideas would have on subsequent research in biology in Australian institutions, even while many Australians were slow to accept them.

The influence of the Enlightenment on patterns of scientific research in Australia and the drive to become a self sufficient nation, viewed as an experiment of the Enlightenment, are a theme of this book. Natural history was the showpiece of the Enlightenment and Joseph Banks its high priest. Too often early science related to Australia is dismissed as being nothing more than Banks' supervision of collectors and catalogues – collectors building collections for the British public and its natural philosophers (including of course, Banks himself). A preferable view is that Terra Australis was part of the science equation, initially as a natural environment of opportunity that powered knowledge and careers like his, and later by accommodating a "can do" population without traditional restraints and regulations. An early example was the botanist Robert Brown who spent four years in the Australian landscape before returning to London and fame, built on his botanic discoveries and ideas they inspired, directly related to his Australian experience. This pattern was to be repeated again and again, whether the scientist in question stayed in Australia, or returned to Europe. For those who came

to call Australia home, innovation and initiative in the use of science to analyse and resolve problems, was the common denominator.

The Banks Era (1770–1820)

Great excitement attended the establishment of a convict settlement in New South Wales, not because England had found a solution to the problem of overcrowded gaols that followed loss of her American colonies, but in anticipation of discovery of new flora and fauna. This was the time of the Enlightenment. The collections of weird specimens and drawings of William Dampier from the west coast in 1699 and those of Joseph Banks from the east coast in 1770, promised vast new information and research opportunities for those who dared. And Banks was the undisputed king of everything Terra Australis, no longer incognita.

Few major decisions of any nature were made without his sanction. He was the catalyst that made things happen. He employed botanist-collectors in New South Wales and rubber-stamped the appointment of others. All specimens would come to his London home in Soho Square, where he and his skilled assistants would collate, classify, study and distribute to the Chelsea Physic Garden and the newly established botanical gardens at Kew. He was much involved in the establishment at Kew in 1772 through the merging of royal estates. The constant supply of exotic new species from Australia helped turn Soho Square into the hub of world botany. Banks' retention of the presidency of the Royal Society for an unprecedented 41 years until his death in 1820 reflected his standing.

The Enlightenment and the associated Industrial Revolution combined to create an affluent and inquisitive middle class. Natural history was the go to area in science that could be easily understood, collected and even grown in the backyard. This provided opportunity for the entrepreneur who saw the need for a literature to support public interest. New magazines combined science with practical information on how to grow the plants. First amongst many was *The Botanical Magazine* published by William Curtis in serial form from 1787, including two or three hand-coloured engravings with each edition. Each plant came with detailed comment

NOUVELLE-HOLLANDE: Île Bernier.

KANGUROO À BANDES. *(Kanguru Fasciatus N.)*

Kanguroo à Bandes in *Voyage de Decouvertes*, Lesueur et Petit (1824)

The Baudin voyage (1800–03) combined a search for La Perouse with serious scientific intent, with a team of natural historians. Most of the published illustrations were by Lesueur. The "Kanguroo" and the "Platypus" would make the most dramatic impression on Europeans.

on its botany and the history of its discovery. The success of this publication at a time when exotic flora from Australia appeared in the Royal Botanic Gardens at Kew and in London's commercial nurseries, was not a coincidence. By the time of Banks' death in 1820, 155 Australian species had appeared in *The Botanical Magazine*, or seven per cent of all plants illustrated by that time. Publication continues today as an official production of Kew Gardens. Most of the early illustrations in *The Botanical Magazine* were by Sydenham Edwards who would later publish his own *Edwards's Botanical Register*, which also contained many "Australian exotics".

An appendix in 1839 to the first twenty three volumes of *Edwards's Botanical Register* includes "A Sketch of the Vegetation of the Swan River Colony" by John Lindley,

professor of botany at University College, London. The appendix included a summary of all published accounts to the time, with a herbarium of an additional 1,000 specimens. Lindley's purpose was to complete documentation, to standardise nomenclature and, as he wrote, to assist residents of the colony "to know what [specimens] to send home". The out dated and patronising view of naturalists who never travelled to Australia, that Australians contributed little science but rather they were collectors for the real scientist back in England, was and remains popular, not understanding that residents were becoming aware of an Australian identity shaped by unique and challenging experiences while adapting to their adopted land. They developed an aspiration to be in charge of their science, working with European colleagues on an equal footing, while retaining respect for European experience and academic skills. This influence would impinge on botany research even after Federation.

A less obvious influence was the impact of being in Australian environments on English scientists who came for a limited time. Perhaps the best example was Robert Brown whose four years in Australia profoundly affected his ideas, and his subsequent stature as the most respected botanist of his time. Brown was the first great Australian scientist, benefitting from unaffected opportunity bathed in the spirit of the Enlightenment.

Botanist David Burton was the first Banks Collector. He arrived in 1791, and after just three months had sent a large collection of seeds and plants back to England before accidentally shooting himself the following year, a not uncommon fate of early botanists.

Others whom Banks commissioned include the chief surgeon with the First Fleet, John White, George Caley, Robert Brown, Ferdinand Bauer, George Suttor, and finally Allan Cunningham, who would spend much of his remaining life in Australia and New Zealand. A highly professional and competent botanist, Cunningham was a watershed marker representing the beginning of the Australian resident scientist.

Joseph Banks recognised the scientific and economic value of colonial botanic gardens, if only to facilitate the transit of specimens back to London and "his" Kew Gardens. During his time, botanic gardens were founded in Sydney (1816) and Hobart (1818). His legacy was an understanding of the culture of science in relation to the remarkable opportunities for natural history research in colonial Australia. What perhaps is less obvious, is his impact on patterns of science that evolved over the century that followed his death, and on into modern times. While James Cook literally put Australia on the map, Joseph Banks put a footprint on the map of Australian science.

Colonial Natural History: Science Finds its Own Feet (1820–1900)

After the death of Joseph Banks, botanical science was closely associated with the development of botanic gardens. The fortunes of early Australian botanic gardens were tied to leadership from Kew. For a long time however there was little dynamic activity by the directors of Australian botanic gardens to encourage the continued aptitude of English botanists. In the immediate post-Banks period botanists at Australian gardens restricted their activities to collecting and classifying new species. By mid-century there had been no attempt to update the comprehensive data set of Australian flora since Robert Brown's epic *Prodromus Florae Novae Hollandiae* in 1810. With Banks gone a royal commission in England was established when it became clear that the Gardens at Kew were losing their gloss. The findings of the commission brought recognition of the imperial importance of science, and the value of the nexus between Kew and colonial gardens. This recognition licensed the father and son directors at Kew, successively William and Joseph Hooker, to support development of botany internationally. By 1900 there would be thirty botanical gardens through the British Empire, with six in Australia.

Between 1863 and 1878 an English natural historian, George Bentham, initially commissioned by the Hookers, worked to update the Australian herbarium as part of the "Kew Series". Any doubts that control of Australian botanical science had shifted to Australia were settled by the arrival in Australia of Ferdinand von Mueller from Rostock, Germany, ostensibly for his health, in 1847. His indispensable involvement with Bentham in the creation of *Flora Australiensis* demonstrated the "change in guard" with crystal clarity. Von Mueller's years were a defining time in Australian botany. With an immense work output he established an Australian credibility amongst his European colleagues. The Kew team torpedoed close collaboration planned between Bentham and von Mueller, despite the fact that von Mueller would contribute his extensive Melbourne herbarium to update Bentham's publication. He went on to add 520 new species in his *Systemic Census of Australian Plants* published in 1882, amongst numerous additional works.

Von Mueller used a modification of Jussieu's natural system of classification. It was then adopted by Charles Moore, long standing director of the Sydney Botanical Gardens, in his comprehensive *Handbook of the Flora of*

"Map of the Country between Bathurst and Liverpool Plains" in *Geographical Memoirs on New South Wales*, ed. B. Field (1825)

This map of discovery by Cunningham illustrates his paper presented to the Philosophical Society, demonstrating a range of skills needed to enable him to achieve his primary goal of collecting and describing the flora of Terra Australis.

New South Wales, published towards the end of his career as Government Botanist. Von Mueller was a somewhat distracted academic with little interest in the social side of running the Melbourne Botanic Gardens.

Charles Moore was a quality manager, quiet and not so well recognised, whose renown has suffered, somewhat unfairly by contrast with Joseph Maiden who followed him as Director. Maiden's prodigious workload, and organised and talented contributions

would leave a permanent mark on the history of the Sydney Botanic Gardens, through a directorship from 1896 to 1924. Maiden was a man of his times, recognising and advancing economic botany. The story of his career makes the perfect bookend to a century of botanical science characterised by the documentation and classification of novel flora. He actively participated in the intellectual life of Sydney and was an important member of the Royal Society of New South Wales, responsible for writing its history.

Post-Federation Botanical Science (1900–50): Taxonomy Replaced by Ecology and Plant Physiology

At Federation, there was no chair of botany. Botany had been taught in all the universities with a focus on taxonomy that reflected interests of the day and in that sense was at the cutting edge. A chair of botany was established in Melbourne in 1906 and held by Professor Alfred Ewart (1872–1937). Though trained in England in plant physiology with a research interest in protoplasmic streaming, Ewart inherited the mantle of pragmatism, with the Victorian Government seeming not to have learnt from its experience with von Mueller and his successors that science and administration do not necessarily go together. He struggled with a high teaching load and responsibilities split each day between the Botanic Gardens and the University – much of each day was spent on the road. He had no time for research on plant physiology, so he adopted the established colonial pursuit of taxonomy and agriculture. He even wrote a book on weeds. Despite the pressures, he published over 150 scientific papers and completed a *Flora of Victoria* as well as field studies and forestry initiatives – an impressive effort. He established botany as a significant scientific discipline at the university, something not achieved by Frederick McCoy and Baldwin Spencer who had preceded him in charge of the natural sciences in Melbourne, but whose interests hardly included botany.

Ewart was followed by Professor John Turner (1908–91), who held the chair from 1938 to 1973. Building on what was essentially a teaching department, Turner promoted basic research in plant physiology and ecology, while continuing attention to economic botany. In World War II, like many Australian research workers, he switched attention to the war effort, studying penicillin producing moulds. With the CSIR and Professor Robertson at Sydney University, he became part of the Plant Physiology Network, establishing field studies in water catchment areas and forest reserves, as well as vegetation studies in high country pastures to monitor the impact of grazing, fifty years before governments sponsored centres of ecological research and land and water conservation.

The second chair in botany was in Adelaide. Established in 1912, it was held by Professor T.G.B. Osborn. It is no surprise that his research interest was strongly influenced by economic issues. He paid particular attention to plant pathology and agricultural botany. He was instrumental in recognising the need for a dedicated research institute in agriculture, and promoted the development of the Waite Agricultural Research Institute on part of a property donated to the University in the early 1920s. His lasting research initiative was to establish the Koonamore Research Site for ecological and revegetation studies in arid conditions, 650 kilometres north of Adelaide. This important programme continues today. When he moved to take the chair of botany in Sydney in 1927, his position was taken by Professor J. Wood, a local graduate who continued his ecological programmes.

Abercrombie Lawson (1870–1927) became foundation Professor of Botany at Sydney in 1913. Lawson like Osbourne, was recruited overseas, and a period began of outstanding leadership in Sydney by academic botanists from North America or England. He was succeeded first by Osbourne (1927–30) and then Professor Eric Ashby whose interests were economic botany and critical thinking and problem solving. They were followed by Professor Rutherford Robertson who established plant physiology at an international level.

Until Lawson came, botany in New South Wales was descriptive taxonomy, and economic – all important and geared to local needs, but out of step with international focus on plant physiology and the impact created by Darwin's natural selection. He introduced contemporary science into teaching and research. Lawson's particular interest was comparative morphology, using as his model, the evolution of gymnosperms, which are non-flowering, seed bearing species such as conifers. He studied Australian gymnosperms.

By 1950, academic departments were teaching and researching many aspects of plant physiology and biochemistry but the most significant contributions were still in economic botany especially in agricultural research institutes including those of the CSIRO. The most important discoveries came from the work of Rutherford Robertson and his students on the stoichiometric and structural basis of ATP synthase and the coupling of proton gradients to the formation of high energy phosphate bonds. The other area of particular importance was in ecological studies in the Australian environment, highlighted by the American-Australian Scientific Expedition to Arnhem Land in 1948 led by Charles Mountford, with its focus on ecology and environmental relationships.

Important Botanists

Robert Brown (1773–1858)

Robert Brown's four years in Australia between 1801 and 1805 established an academic credibility for colonial science, and a reputation for himself as one of the leading scientists of his times. His studies of morphology and geographic distribution of Australian proteaceae shook contemporary thinking. Brown came to Australia as one of Banks' collectors, but he was much more than that. He left London on the pay of the Irish army, and returned as botanist in charge of Banks' collection, soon to become one of the world's most respected botanists following publication of his Australian studies after accompanying Matthew Flinders on his circumnavigation of Australia. Brown spent months in Tasmania and New South Wales. As a taxonomist he switched to the natural markers of Jussieu, but it was his farsighted ideas from keen observation of Australian flora in situ that brought him eminence. Many of his ideas are expressed in his appendix to Matthew Flinders' journals, published in 1814. In a broad overview of plant distribution, he used ratios of dicotyledonous to monocotyledenous plants to assess plant geography, remarking on differences between Australia and the northern hemisphere that he could not easily explain in terms of soil or climate. These observations anticipated both ecology as a science of great importance in an Australian context, and the differential import of micro-deficiencies based on physiological differences between monocots and dicots in retention of essential nutrients. Perhaps Brown's most intuitive observation regarded distribution patterns of proteaceae species. He described a distinct difference between species on the east and west coasts but noted similarities between those on the east coast with species in America and those on the west with examples in Africa. This remarkable observation predicted the idea of continental drift advanced by Alfred Wegener a century later.

His reputation was made by his monumental *Prodromus Florae Novae Hollandiae et Insulae Van Diemen*, published in 1810 and updated twenty years later to include additional proteaceae collected by others including Allan Cunningham. The Prodromus was the most complete record of Australian systemic botany to that time, and indeed remained the reference collection until the update by George Bentham and Ferdinand von Mueller in 1868–78. Brown never again travelled to Australia but kept in contact with botanists and explorers – his last written contribution being in the appendix to Charles Sturt's *Narrative of an Expedition into Central Australia* (London, 1849) – in which he neatly summarised the position of Australian botany mid-19th century. He acknowledged an involvement with the herbaria of Allan Cunningham (New South Wales and the northwest coast), William Banter, James Drummond and M. Preiss (Western Australia), and Ronald Gunn (Tasmania), extending the number of species known in Australia to about 7,000. He reviewed about 750 species collected in "interior Australia" by Sturt, Mitchell and Oxley. He considered that plant life in those areas resembled plant life in southern parts of Australia explored earlier on, but fewer "tribes" otherwise characteristic of better known parts, and lacking many unique species characteristic of

what he called the "principal parallel" between 33°S and 35°S. His use of physiological classification, enabling comparison between species, and his keen perception of plant distribution, link him forever to the two great discoveries of natural selection and plate tectonics.

Allan Cunningham (1791–1839)

Allan Cunningham broke the mould – he was the first professional botanist of the Banks-Brown era to call Australia home. He spent most of the second half of his life in Australia. His passion for scientific botany and discovery of new species led him to accompany John Oxley on Oxley's expedition along west flowing rivers seeking answer to the question of whether there was an inland sea. He also went with Phillip Parker King, charting the northern coasts of Australia with the hope of finding the great river which drained an inland sea. Cunningham led his own scientific expeditions, discovering along the way important pasture in the Liverpool Plains and the Darling Downs. For a short time in charge of the Botanical Gardens in Sydney, he made his vision clear to Governor Gipps of a Kew Garden in the South: to "create a depository of every species" and to acclimatise every economic plant in the region, with a seed depot for international exchange, a vision never to eventuate.

Consistently with the times, economic botany was always dominant in Cunningham's plan (provided he did not have "to provide cabbage to the Governor" – a reference to his distasteful ex officio function of caring for the Government House vegetable garden).

Allan Cunningham was a watershed scientist – never sure where his priorities lay, but with a vision of what science could become in Australia. His life was too short owing to tuberculosis, and his potential not fully realised.

Barron Field (1786–1846)
Alexander Macleay (1767–1848)

Barron Field and Alexander Macleay were not scientists in a traditional sense. However, they were passionate sons of the Enlightenment who on the one hand reflected the impact of the Enlightenment on men of influence, and on the other, used their positions to promote science

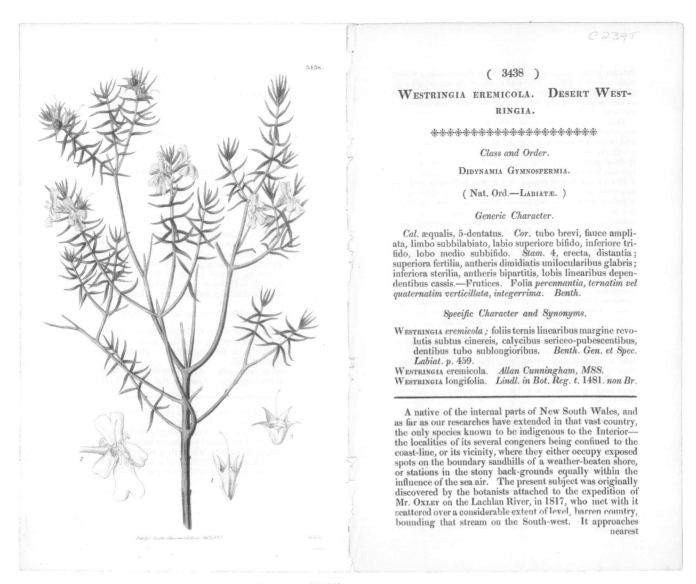

Westringia eremicola in Curtis's *Botanical Magazine* (3438)

within the fledgling colony. Though very different personalities who never met, their contributions to science were polarised around a scientist of significance, the controversial Governor Thomas Brisbane.

Barron Field was an ambitious judge, whose seven years in Sydney were mired by conflict and bias. He was something of a polymath, active in Sydney's artistic and scientific circles. His singular lasting contribution to science was editing *Geographical Memoirs on New South Wales by Various Hands* (London, 1825). He was a political survivor. Despite negative reports from Governors Macquarie and Brisbane, and from Commissioner John Bigge (sent to the colony to report to London regarding complaints about Macquarie's administration), he remained well regarded in the Colonial Office. His

position in Sydney was terminated following the Bigge Report but was followed by appointment as Chief Justice in Gibraltar, a position he held until retirement in 1841.

The *Geographical Memoirs* included the first set of publications from the Philosophical Society of Australia which "soon expired in the baneful atmosphere of distracted politics". He anticipated "our little society will be resuscitated by the new colonial government," which included "the natural science and business talent of the Colonial Secretary," who was Alexander Macleay. The contents included summaries of the current state of knowledge regarding astronomy, botany, geology (by Alexander Berry – a prominent businessman) and meteorology (by Governor Brisbane). It is of interest that in this collection of reports on science, there are

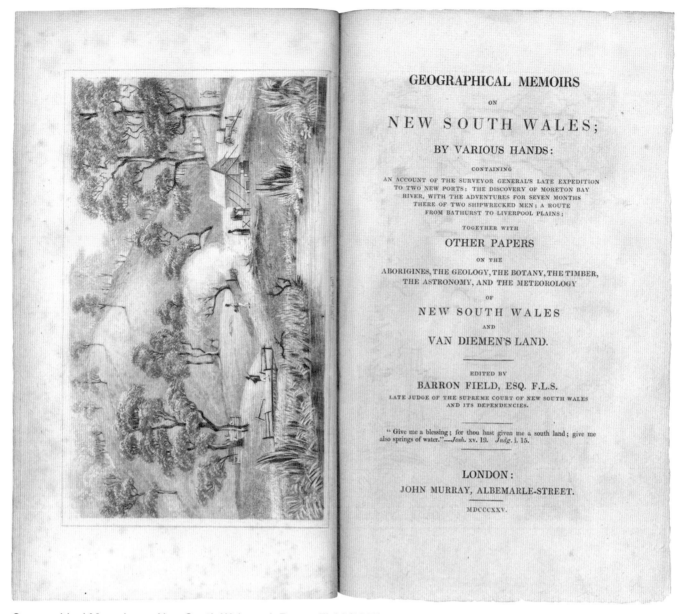

Geographical Memoirs on New South Wales, ed. Barron Field (1825)

important contributions by John Oxley, Allan Cunningham and Phillip Parker King on discovery and exploration, Aborigines (by Barron Field) and economic botany. Anticipating the broad interests of the Royal Society of New South Wales that would follow, Field includes *The First Fruits of Australian Poetry* – which of course were largely his own work!

Alexander Macleay (also spelt McLeay) was recruited at age fifty eight to help Governor Darling restructure governance in New South Wales, after the term of Governor Thomas Brisbane, who was accused of much too much science and not enough administration. Pressured into coordinating and supervising administration in colonial New South Wales, Macleay accepted the position mainly to expand his natural history collection. At that time his collection of insects was probably the largest in the world, and his love of science combined with obvious administration talents, had led to senior positions in the Royal Society and the Linnean Society in London. As with Barron Field, Macleay did not avoid controversy, but worked tirelessly to manage the affairs of the colony in support of successive governors. He took leadership in community science supporting the development of Australia's first

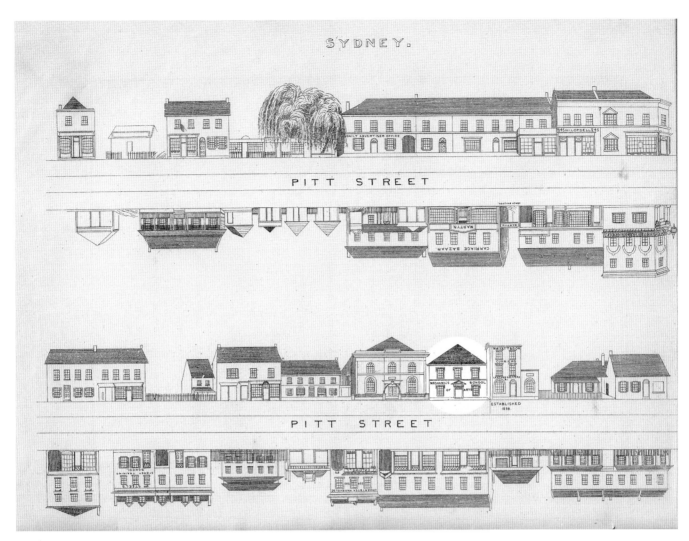

Sydney in 1848, Joseph Fowles (1849)

In the fledgling colony at Sydney Cove two organisations supported and promoted science — the Philosophical Society of Australia (precursor to the Royal Society of New South Wales) and the Sydney Mechanical School of Arts. Both were inspired by the Scottish Enlightenment. Barron Field's *Geographical Memoirs* includes five papers on scientific subjects presented to the Philosophical Society, and Fowles's profile of Pitt Street shows established accommodation for the Mechanical School of Arts.

museum, the Free Lending Library and the local chapter of the Linnean Society. As chairman of a committee to oversee the development of the Botanical Gardens and the Museum, he began an association with Sydney's scientific community which would continue through a family dynasty for the rest of the century.

His eldest son William Sharp Macleay, who arrived in 1839 inherited his father's collection and interest in museums. After his death the collection passed to his cousin William John Macleay (1820–91) whose role as trustee at the Australian Museum sparked controversy around the dismissal of Johann Krefft. Cousin William

organised natural history collection expeditions especially to the Torres Strait and New Guinea – to enlarge the Macleay collection, which he donated to the University of Sydney to form the important Macleay Museum of natural history.

The Macleay family did much to support and promote natural history in their adopted home, through their association with the Museum, involvement and donations to the Linnean Society, and in direct help to individual scientists. Perhaps the most important support was to Nikolai Miklouho-Maclay, the Russian biologist, ethnologist and anthropologist who did foundation studies around Australia and especially in New Guinea. Sydney became a base for Miklouho-Maclay, in part because he married the daughter of Sir John Robertson, a Premier of New South Wales.

Barron Field and the Macleay family represent an important segment of amateurs of science whose support and administrative contributions provided a backbone to the fledgling scientific community in colonial New South Wales over an eighty year period. Controversy and even competition with public collections diminish, but do not erase, the importance of their roles in establishing infrastructure and communication in Colonial times.

Ferdinand von Mueller (1825–96)
Joseph Maiden (1859–1925)

Two botanists, the one from Germany, the other England, with responsibilities rooted in colonial botanical gardens, both leading scientists of colonial Australia with enormous work capacity, were very different people. They represent the pinnacle of descriptive botany and promoters of economic botany in late Colonial times. For von Mueller the Melbourne Botanic Gardens was a base and a laboratory, while Maiden focussed his attention on the Sydney Botanical Garden recognising his broader role in public education perhaps because of his background in museum management.

Ferdinand von Mueller, a botanist and pharmacist, migrated to Adelaide to find new opportunities in a healthy climate. In 1852 he was appointed Government Botanist of Victoria, with responsibility for the Botanic Gardens. He immediately set about collecting and classifying plants in Victoria and the rest of Australia. He sent duplicates to Kew Gardens, expanding the Herbarium of Australian plants, a collection that was the basis of his collaborative publication with George Bentham, *Flora Australiensis*. He was botanist with the North Australian Exploring Expedition led by A.C. Gregory in 1855–56 from Victoria River to Moreton Bay – a trek of 5,000 miles during which he collected nearly 2,000 species, half of which had not been described. His scientific publication rate was extraordinary – including over 800 papers with his monumental *Fragmenta Phytographica Australiae*, published in twelve parts between 1858 and 1882. Although he

was a prominent figure on the local stage, he maintained a wide circle of European contacts, important for the acceptance of an emerging science out of Australia.

Joseph Maiden was thirty five years junior to von Mueller, but in many ways led a parallel life: trained in science, with a longstanding love of natural history, he migrated to Australia for health and opportunity. Through contact with Professor Liversidge – that great catalyst of so much science in late colonial New South Wales – Maiden became curator of the new Technology Museum. After destruction of the Museum and its collections in the Garden Palace fire of 1882, he supervised its move to Ultimo. His passion, however, was botany and when Charles Moore retired as Director of the Botanical Gardens in 1896, Maiden took the role of Government Botanist and Director of the Botanical Gardens, the New South Wales counterpart of the position once held in Melbourne by von Mueller. In an academic sense his output was little short of that of his Victorian forerunner. He presented forty five papers to the Royal Society of New South Wales, His interest in taxonomy continued the theme of colonial Australian botany with productions such as *A Critical Revision of the Genus Eucalyptus* published in seventy parts from 1903, which included 366 new species, and *Forest Flora of New South Wales* in seventy seven parts from 1904.

These two European born botanists dominated botanical science in Australia for nearly eighty years. Aware of treading a line between local and international science, and seeing themselves always as Australian, they established Australia as a leader in international descriptive botany. Von Mueller's energetic reaction to George Bentham's attempt from his office to take leadership of Australian botany in Kew places von Mueller foremost in defining Australian science, along with William Clarke who resisted the dating of the Newcastle coal seams as Jurassic (like European coal seams), by the English school of Sedgwick and McCoy.

Maiden and von Mueller were sensitive to the needs of the emerging nation through economic botany including forestry. At a time when most influential people believed trees exist to be used, both had strong views on forest renewal and conservation. They stayed in touch with international botany through personal

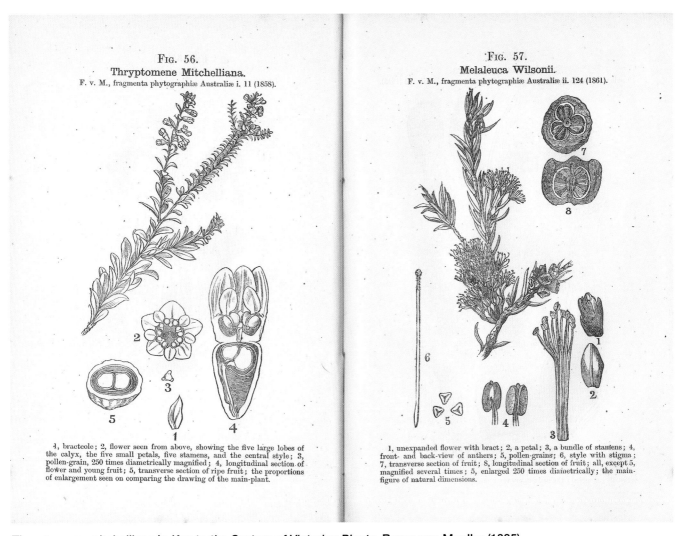

FIG. 56.
Thryptomene Mitchelliana.
F. v. M., fragmenta phytographiæ Australiæ i. 11 (1858).

FIG. 57.
Melaleuca Wilsonii.
F. v. M., fragmenta phytographiæ Australiæ ii. 124 (1861).

1, bracteole; 2, flower seen from above, showing the five large lobes of the calyx, the five small petals, five stamens, and the central style; 3, pollen-grain, 250 times diametrically magnified; 4, longitudinal section of flower and young fruit; 5, transverse section of ripe fruit; the proportions of enlargement seen on comparing the drawing of the main-plant.

1, unexpanded flower with bract; 2, a petal; 3, a bundle of stamens; 4, front- and back-view of anthers; 5, pollen-grains; 6, style with stigma; 7, transverse section of fruit; 8, longitudinal section of fruit; all, except 5, magnified several times; 5, enlarged 250 times diametrically; the main-figure of natural dimensions.

Thryptomene mitchelliana in *Key to the System of Victorian Plants*, Baron von Mueller (1885)

This species was first identified by von Mueller in 1858, and was included in his encyclopaedic *System* of native Victorian plants. It is typical of his scholarly and detailed approach to the documentation of Australian botany, and of his connecting of Australian science with international interests.

contact, publications and plant and seed exchange programmes.

Both were extensively involved in local science organisations and comparable in their commitment to science, but they differed in their commitment to getting the public interested in science. Von Mueller had little interest in organising attractive arrangements and structures for the public (and correspondingly lost control of the Gardens after twelve years). Maiden had a clear idea of what was expected, communicating with the public, extending his influence to local parks and streetscapes, while producing education programmes for an interested public. Von Mueller and Maiden were the face of Australian descriptive botany and themselves

a link to commercial botany and the watershed with plant ecology and physiology then developing in the new universities.

Theodore Osborn (1887–1973)

T.G.B. Osborn was recruited to the Foundation Chair of Botany at the University of Adelaide at the age of twenty five with Classical training and experience in taxonomy, plant pathology and agricultural botany. Like earlier botanists in Australia Theodore Osborn represents a watershed in science. In his case an academic switch to study ecology in an Australian setting. Appointed in 1912, Osborn immediately recognised the importance of using the principles of ecology – then emerging as a science of its own – in an Australian environment. Initially he addressed issues in forests to the immediate north of Adelaide and then the very practical problems of grazing in semi-arid regions, at a time when overgrazing was not recognised. The reality was that the great saltbush and bluebush plains faced becoming plantless deserts.

Osborn recognised the value of rural laboratories. Graziers made land available from a station 400 km northeast of Adelaide, the Koonamore Vegetation Reserve, where an unreliable rainfall averaged around 20 cm per year. This programme continues after ninety years, as the world's longest continuous ecological programme in arid country. It emphasises the importance of long-term study in arid zone ecology where low germination rates occur due to variable rainfall.

Osborn was one of the first academics to recognise the value of separate research institutes where major committed focus can be applied to solve problems and advance science. He was the driving force in establishing the Waite Agriculture Research Institute in 1924 on the Urrbroe property, donated to the University by Peter Waite.

In 1927 he moved to Sydney University where he extended his research interests to coastal vegetation. Throughout he worked on practical problems with the CSIR. Being recruited back to the Sherardian Chair of Botany at Oxford was a tribute to his Australian research. Following retirement, he moved back to Adelaide. An inspiring teacher, he influenced many young Australians, including his successors at Adelaide Professors J.G. Wood and Rutherford Robertson, and his legacy, the Koonamore Vegetation Programme.

Rutherford Robertson (1913–2001)

Rutherford Robertson's contribution as a botanical chemist was judged by some as being worthy of sharing the Nobel Prize 1978 awarded to Peter Mitchell for ion transport and energy capture in mitochondria based on the chemiosmotic theory. Robertson's research focused on respiration in plants and its link to energy production through phosphorylation and creation of high energy phosphate bonds in ATP. He maintained his laboratory science throughout a long and impressive career. Guided by Professor Osborn in Adelaide he chose plant physiology, working on mechanisms of opening and shutting stomata on leaves. He was awarded a Linnean Society scholarship to work in Sydney and there obtained an 1851 Exhibition Scholarship to work with Professor Biggs in Cambridge, on active transport systems in plants – a short jump to mitochondria ion transport and energy generation. Returning to Sydney as war broke out, he switched to research practical issues of importance such as how to prolong storage time of apples, and what caused heat to be generated in stored wheat. This brought him in close contact with the CSIR, which would become an important part of his professional life, including a senior appointment that allowed both basic and applied research. He would take most of the more senior positions in science available in those times – Professor of Botany at Adelaide and Sydney Universities, the Executive of CSIRO and finally Director of a Research Institute in the Australian National University. His teaching research and administration skills and achievements merged in a way few have managed. His work effort, prestigious positions, and personality made him an influential role model for many students and colleagues.

Zoology:
Better Late Than Never

From the time of William Dampier's visit to the western shores of Australia, Australian animals, from beetles to kangaroos, fascinated Europeans.

Collections were made for study in England and the first museum in Sydney (1825) was to store specimens for English interest. The collection made by Dampier at Shark Bay is now kept in Oxford. The journals written by the First Fleet authors and explorers documented a fauna unknown to the western world. However as Barron Field noted in *Geographical Memoirs* in 1825, zoology fell behind studies of flora and indeed other disciplines of science. Field expressed hope that the arrival of Alexander Macleay the following year, as Colonial Secretary for New South Wales, with his world-best collection of insects, would be a catalyst of change. Macleay came, and did much to stir interest in science, especially zoology, and his great collection – after being expanded and broadened by his nephew – would be donated in 1890 to the University of Sydney as the foundation of the Macleay Museum. Fascination, however, did not lead to scientific research of any real consequence in zoology, other than occasional attempts to catalogue collections by such as Professor George Bennett when he was Director of the Australian Museum in Sydney.

There was little change until the appointment of a German naturalist, Gerard Krefft, as curator at the Australian Museum in 1861. Krefft had accompanied fellow German Wilhelm Blandowski (1822–78) on a collecting trip in 1857 around the junction of the Darling and Murray Rivers. Krefft was then employed cataloguing specimens in Melbourne. His cataloguing of specimens stored at the Museum in Sydney plus his papers on many aspects of zoology established standards of scientific enquiry that were new to Australia, standing out in a sea of amateur interest. Unfortunately, the Museum was controlled by amateur "collectors" opposed to Krefft and what he was trying to establish.

"A Fish taken on the Coast of New Holland etc" in *A Voyage to New Holland &c in the year 1699* (vol III). Captain William Dampier (1703)

Dampier's book demanded attention — in part because it presented a "new" biology to the English public.

Krefft was illegally dismissed by the trustees in what was a major scandal, made worse by his unorthodox behaviour. Zoology's dependence on Krefft can be seen in a review of articles published by the Royal Society of New South Wales. Krefft was one of the main contributors, but following his departure, only one per cent of articles addressed a zoological topic.

In the latter part of the century, the *Proceedings of the Linnean Society* became the favoured journal in Australia for zoological contributions, where they comprised about one third of published articles, on topics ranging from insects to monotremes. They were usually of a descriptive or taxonomic nature, submitted by museum staff, interested amateurs, and increasingly by academics from across the colonies.

Blandowski shared a similar story to that of his younger companion Gerard Krefft. Their expedition to the junction of the Darling and Murray Rivers brought back 17,400 specimens in twenty eight boxes. His personality clash with the Professor of Natural Science and Museum Director has an eerie resemblance to the

"Birds of the Arnhem Land Exhibition" in *Records of the American Australian Scientific Expedition to Arnhem Land* (vol 4, 1963)

This last report published from the expedition records 191 species of birds, adding important new data on locality, variations, breeding and migration.

"natural history" meant they had to teach the disciplines of botany, zoology, geology and often chemistry! Zoology was not a strong suit of most early appointees. The first specialist chair in zoology was Wilfred Agar, appointed in Melbourne in 1919. During his long tenure (1919–48) Agar struggled to find time for research, though he supported academic aspects of zoology and introduced new areas such as cytology and genetics. He did develop a research interest in genetic traits of cattle, which was a precursor to the use of genetic screening in cattle breeding – a continuation of focus on economic science from Colonial times.

There was a common academic interest in museums as a source of material to study and teach in Melbourne. One of the first acts of Fredrick McCoy, Melbourne's new Professor of Natural Science (1854–99), was to commandeer the primitive colonial museum and site it in University grounds. McCoy was a geologist more than a biologist, whose long and tenuous career in Melbourne was never without controversy. In Sydney the controversy was more about the Macleay family than the world class collection they donated to Sydney University in 1890, while in Adelaide the Tate Museum was assembled by Ralph Tate, integrated into the life of the University, and seen as a legacy of a favourite son of Adelaide University.

In New South Wales, the Zoological Society brought together those interested in research. Its start was not impressive, despite being organised around Professor George Bennett (Australian Museum), Charles Moore (Botanical Gardens) and Henry Parkes (the influential politician). Science was always in conflict with practical aspects of zoo management. After a period between 1861 and 1879 when the society became the Acclimatisation Society of New South Wales, the organisation was reshaped as the New South Wales Zoological Society. The main goal between 1881 and 1916 was to establish and run a zoo in Moore Park. It was not until the Taronga Park Trust was established and the zoo was moved to Taronga Park that science and research finally became a priority.

The key to research was *The Australian Zoologist* published between 1914 and 1920, with contributions from museums and universities across the country.

later conflict of Gerard Krefft with William Macleay. However, Blandowski lacked the scientific enquiry and skills of Krefft, and he returned to Germany and obscurity.

A new era for zoology came with the establishment of universities and their chairs of natural history at Sydney, Melbourne and Adelaide. Those appointed often had a record and enthusiasm for research, but adjusting to a new culture, building a department and curriculum from scratch, left little time. A major challenge was that

"Great Barrier Reef Corals Plate IV" in *The Great Barrier Reef of Australia*, W. Saville-Kent (1893)

Saville-Kent, President of the Royal Society of Queensland and Head of Government Fisheries, presented the most up to date summary of corals and other constituents of the Great Barrier Reef, interestingly commenting on contemporary bleaching and discussing the reef in the commercial framework that framed science in colonial Australia.

W. Saville-Kent, del. et pinx. ad nat.   Riddle & Couchman, imp. London, S.E.

GREAT BARRIER REEF CORALS.

Examples from the list of contents include: H. Burrell on monotremes; W. Dakin on marine zone ecology; W. Froggatt on entomology and ornithology; L. Howeson on evolutionary biology; L. Le Souef on mammals; A. McCullough on icthyology; T. Johnston on parasitology; W. Rainbow on entomology; and Professor of Zoology T. Flynn (best known as Errol's father) on marsupial embryology.

The Challis Chair for Zoology at Sydney University was initially given to William Haswell as the foundation professor of biology, changing to zoology in 1913. Haswell had come to Australia "for his health" and quickly established a reputation amongst those who counted, especially the Macleay family, as an informed enthusiast for research in marine biology. He would have an impact on zoology research until his death in 1925, with a programme in ecology of the continental shelf. He was active in the Zoological Society and as a trustee of the Australian Museum. He also produced an important text on zoology.

Haswell was followed by Lancelot Harrison (1880–1928), an Australian science graduate with a passion for zoology. His long term research into host-parasite relationships involving mallophaga (biting lice) was

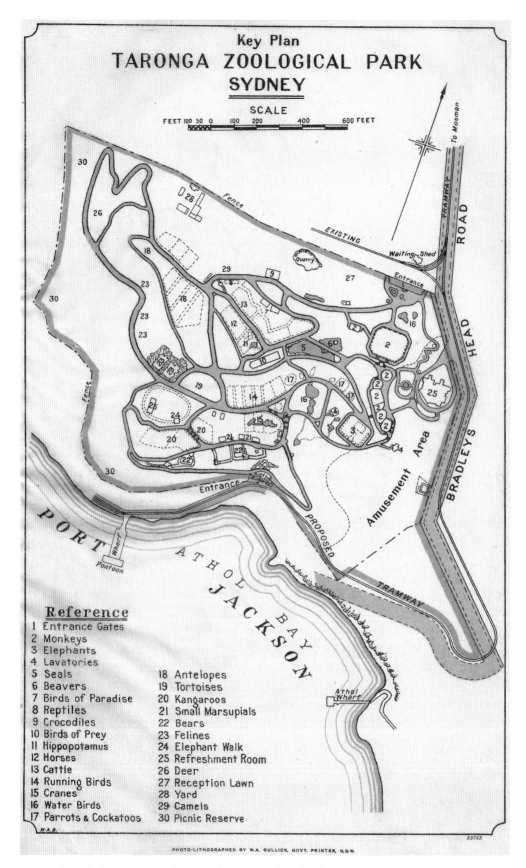

Taronga Zoological Park, in *Report for Year Ended December 1916*

Taronga became Sydney's new zoo when the zoo at Billy Goat Swamp became overcrowded. The Park was supervised by the Zoological Society, its design influenced by a review of the *Zoos of the World*, assessed on a tour made in 1908.

world class, introducing the idea of study of a different form of ecology that would become a major area of Australian research.

William Dakin, who followed Harrison, holding the position from 1929 to 1949 established a research interest in ecology of shorelines. Research shifted to morphogenesis, inflammation and embryology (the latter continued by his successor Patrick Murray who held the Challis Chair from 1949 to 1960). Charles Birch joined the Sydney zoology department in 1948 and would take the Chair in 1960. His early studies in South Australia at the Waite Institute of Agricultural Science changed thinking in ecology by challenging dogma on self regulation of populations through competition for resources. He showed a significant impact of external factors like weather. His later philosophical views on life form and consciousness muddied thinking on natural selection by introducing the idea that all species influence each other in any particular environment.

By the mid-20th century, zoology had moved a long way from classification and phylogeny. Strong research programmes in ecology across environments and species had developed in academic departments while in medical institutes, animal physiology research in areas such as immunology and neurophysiology was leading the world. In university veterinary departments and the CSIRO, animal pathology as well as physiology was studied with particular strengths in the investigation of causes and management of pathogenic host-parasite relationships.

Important Zoologists

William Macleay (1820–91)
Frederick McCoy (1817–99)
Ralph Tate (1840–1901)
Gerard Krefft (1830–91)

The natural history museum was a focal point in zoology in colonial Australia. The idea of a museum in colonial Australia was promoted by Alexander Macleay, who arrived in Sydney with Governor Ralph Darling, to be the Colonial Secretary. He was a talented and well recognised amateur naturalist with a large insect collection. Macleay promoted the idea of a colonial museum to the Colonial Office in London, and in 1827 Earl Bathurst directed Governor Darling to provide £200 per year to create a "public museum at New South Wales". The idea was supported by British naturalists wanting specimens of what Phillip Parker King called the "extraordinary assemblage of indigenous productions". They saw the museum as a holding place for collection and preparation of zoological specimens prior to transport to London. In time it would become the Australian Museum and play a central role in Australian natural history research, but over its first forty or so years, the Museum played little part in the scientific life of the colony. There were some notable moments such as when George Bennett was Curator (1835–41) and produced the first catalogue of collected specimens in 1837.

The face of museums in Australia would change essentially as a result of the efforts of four men: William John Macleay, Frederick McCoy, Ralph Tate and Gerard Krefft. William John Macleay was the nephew of Alexander Macleay and inherited Alexander's extraordinary reference collection of insects. He came to science late in life and became an avid collector and a powerful supporter of those scientists he approved of. By donating the combined Macleay collections to the University of Sydney he created a major research resource. Sydney then had three museum collections, (the Australian Museum, the Macleay Museum and a new Technical Museum, initially curated by Joseph Maiden, that would be developed on a site in Ultimo, eventually to become the Powerhouse Museum).

Frederick McCoy and Ralph Tate were the foundation professors of natural history at, respectively, Melbourne and Adelaide Universities. Both had impressive research track records in palaeontology in Britain, and experience with museums. Both had worked with leaders in geology research. McCoy's application was supported by Adam Sedgwick at Cambridge and Tate was recommended by Thomas Huxley (the man known as Darwin's Bulldog for his zealous promotion of evolutionary theory). McCoy and Tate had a broad experience in all aspects of natural history and were expected to teach across the

PLATE VI.

VENOMOUS.

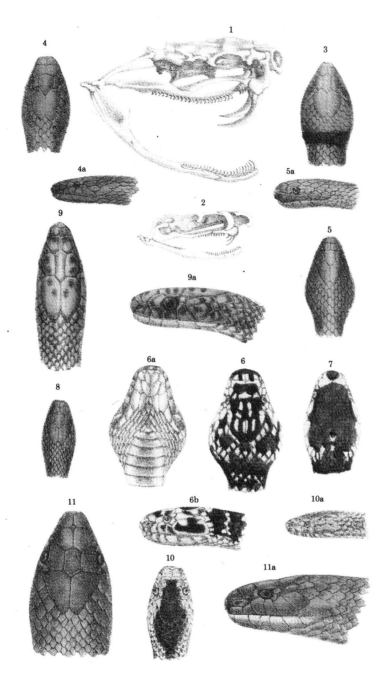

HEADS OF AUSTRALIAN SNAKES.
No. 1. Skull of American Rattle Snake, (Crotalus Durissus.)
No. 2. Skull of Australian Black Snake, (Pseudechis porphyriacus.)

Harriet Scott, del. et lith.   Gibbs, Shallard, & Co., imps.

Heads of Australian Snakes, in *The Snakes of Australia*, Gerard Krefft (1869)

Krefft was the first science based curator of the Australian Museum, bringing a vitality, objectivity and rigour to zoology that had previously been lacking. His main interest was documenting and studying snakes. The illustrator was Harriet Scott (whose work on butterflies is illustrated on p.167). Natural history illustration was amongst the few opportunities in science available to women in colonial Australia.

spectrum, as well as carry out research and promote the discipline.

They both made Australia home while keeping alive their international connections. They both became involved in promoting research at a local level and contributed to economic ventures. That is where the similarity between them ends.

McCoy held Melbourne's chair of natural science from 1854 until his death in 1899. Despite his Irish background and temperament he was much of a classical British mould, conservative and out of touch with "Enlightenment Australia". He was surprisingly an "indoor naturalist" with aptitudes and opinions related to his earlier British experience, rather than openness to opportunities of a new independent world. He was in fact, a throwback to a more rigid and reactionary society.

Nowhere was this more obvious than in opinions McCoy expounded on geology, causing much angst at the time. For example, he denied the presence of deep gold reefs in the goldfields and he maintained a tense argument with William Clarke on the age of the Newcastle coalfields. Clarke's arguments that the coal seams differed from those in Europe, based on stratigraphic studies in the field, were strongly opposed by McCoy who insisted on a British concept of fossil sequence. Clarke proved to be correct but not before unpleasant argument waged through local scientific journals. Controversy was part and parcel of McCoy's life.

What he did well was develop a quality museum, albeit by displacing others and moving it to the University where it remained until after his death. He organised a magnificent collection of natural history artefacts, geological specimens and mining paraphernalia. It was "his" collection. By the 1880s it received about 110,000 visits each year. McCoy's research focus was on classification, and he published catalogues of the fossil collections and zoological specimens. He maintained a botanic garden for teaching, also within the University grounds. Following his death in 1899, his colleague Baldwin Spencer managed the museum, which was moved to a public site in Russell Street. Reflecting his interests, Spencer added important anthropological artefacts.

Ralph Tate joined the younger University of Adelaide in 1875, where he remained until his retirement in 1901. His legacy was very different to McCoy's. While he quietly put together a natural history museum within the University including classical natural history, geology and anthropology, it was integrated into the academic life of the University. It continues today as the Tate Museum. His leadership in science was set against his contributions in geology, zoology and botany. His strong belief in field studies led to pioneering geological research defining tertiary strata in coastal regions, recognising glaciation periods, identifying Cambrian series on the York Peninsula and studying Mesozoic strata in the Great Artesian Basin. The breadth of his contributions to natural history is reflected in his publications in zoology especially regarding mollusca, and in botany. His foundation document of South Australian botany, *Handbook of the Flora of Extratropical South Australia*, was published in 1890. He drove local and national science organisations, becoming president of the Australasian Association for the Advancement of Science, and being awarded the W.B. Clarke medal. He recognised the importance of collaborative interactive science by his participation in major field excursions to the Northern Territory and with the Horn Expedition of 1894 to the Territory's Finke River region. A distinct and forceful personality, he inspired loyalty and cohesion, establishing a platform for science in South Australia, with his natural history museum, a quiet reminder of his extraordinary contributions. Vallance described his contribution on Cambrian fossils as the "emergence of Australian palaeontology from colonial bondage".

While the behemoths of early academic science shaped natural history museums of very different types in Melbourne and Adelaide, it was a self taught non-academic German who would lift the Australian Museum in Sydney from its state of rest. Gerard Krefft came to Australia in 1852 to find gold, and instead became Australia's leading research zoologist. He was also the catalyst – much to his own disadvantage – who initiated serious science programmes at the Museum. In his fourteen years in Sydney, he became internationally recognised for his studies in natural history – ranging from researching snakes to cataloguing the Museum's

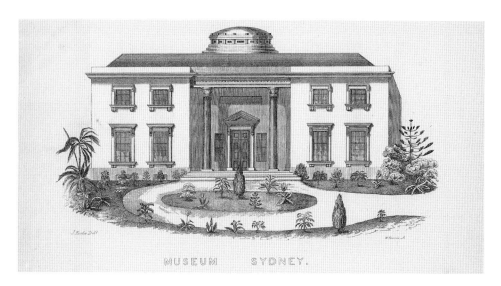

Museum of Sydney, J. Fowles in *Sydney 1848*.

The Museum was established in 1827 to preserve (for English scientists) many "rare and curious specimens of natural history". It had a chequered history until this building was constructed in 1846 and Gerard Krefft was appointed curator in 1861. It became a major site of natural history research, and its collection grew to 21 million cultural and scientific objects.

fossils and minerals. His fieldwork included excavation of fossil deposits in the Wellington Caves.

Krefft was an ardent advocate for the principal of evolution by natural selection. It brought him into conflict with politically powerful opponents, including the Macleay family, who retained the pre-Darwinian idea that evolution was driven by mental and physical strengths.

Krefft was a poor politician, and challenged the trustees for exploiting museum resources (nothing is new) and promoted an independent Australian science platform. It may have been what the Museum needed, but unsurprisingly led to his dismissal.

His stand marked the beginning of independent careers within the museum, and he facilitated career paths for those who followed, including two who would succeed him. First was Robert Etheridge (1866–1920), a palaeontologist of international repute, whose contributions to Australian geology were immense especially in the area of Palaeozoic fossils. He worked with Ralph Tate in South Australia, Alfred Selwyn in the geological survey of Victoria, Charles Wilkinson in the geological survey of New South Wales, and Robert Jack in Queensland. Edgeworth David recognised his contribution to Australian stratigraphy: "the classification and correlation of the coalfields, the goldfields, artesian water basins, oilfields and other mineral deposits of the Commonwealth are based essentially on [Etheridge's] work" Second was Charles Anderson, who followed Etheridge from 1921 to 1940. He was a mineralogist who

switched to vertebrate palaeontology, and continued quality research at the Museum.

Lancelot Harrison (1880–1928)

Lancelot Harrison was a man of energy, with a commitment to academic zoology and a fascination with nature that began in his early life in the Central West of New South Wales. His career followed a classical pathway of that time – a bachelor of science at the University of Sydney followed by the award of an 1851 Exhibition Scholarship to study entomology at Cambridge. A period during World War I working on insect related human disease was followed by appointment in Sydney to the Challis Chair in zoology in 1922. His impact was dynamic, creating a research oriented department as a model for academic development. He was active in every local group pursuing natural history, including the Royal Society of New South Wales, the Royal Zoological Society of New South Wales and the Linnean Society, and he was a trustee of the Australian Museum. His work in entomology was what won him an international reputation, but he was using insects as tools to look at big biological questions. His short working life focussed on host-parasite relationships, especially as they involved mallophaga or biting insects in domestic and wild animals and birds. He examined such relationships to learn whether biting lice evolved in parallel with their bird hosts. He also studied the origins of Australian fauna, research which emphasised his connection with the two "great ideas" of evolution and continental drift.

William Dakin (1883–1950)

William Dakin was an English trained zoologist who followed Lancelot Harrison into Sydney's Challis Chair – which he occupied from 1929 to 1948 – after efforts to pursue academic zoology in Perth. From a wide range of research interests, it would be his passion for the study of Australian seashores that inspired many students into a career in ecology. He continued the tradition of participation in the Linnean Society and the Royal Zoological Society of New South Wales, and of serving as a trustee of the Museum. He stressed the value of applied research in his work on plankton, on the development of a fisheries laboratory and with the CSIRO. He followed Archibald Liversidge's example in contributing to the school curriculum, and delivered radio lectures to the public. His long term studies were included in *Australian Sea Shores* published in 1952, after his death, with considerable help from Isobel Bennett and Elizabeth Pope, who had worked with him for many years. Dakin maintained the momentum of academic zoology in Sydney and kept a focus on ecology relevant to Australian conditions. His major contribution was his work on plankton in Australian waters, though the broad approach he took to marine ecology limited his specific research achievements.

Charles Birch (1918–2009)

Charles Birch was a remarkable Australian born scientist whose career bridged World War II, and who would influence several generations of scientists – role model and enthusiast who addressed important questions in ecology and economic zoology, while bringing new philosophical views to evolution theory and natural selection.

Birch was an undergraduate in Adelaide, and his most important work was done during six years at the Waite Agricultural Research Institute, much influenced by his supervisor H.G. Andreworth, pursuing a research interest in insect ecology. This collaboration culminated in publication of *The Distribution and Abundance of Animals* in 1954. While at the Waite he also addressed economic challenges such as the protection of stored wheat, and the prediction of the path taken by plague grasshoppers. However, his most important contribution to science was a challenge to the accepted dogma in ecology that animal populations are self regulating through competition for resources. He showed that environmental factors such as weather could have a significant impact on controlling the numbers and the distribution of animals.

In 1949 Birch moved to Sydney and in 1960 he took over the Challis chair. In Sydney he focussed on evolution and ecology, motivated in part by his view that all animals have a consciousness, which could have had a bearing on natural selection, by an interspecies effect. He continued his interest in economic ecology with studies of the Queensland fruitfly. Concern about the geographic spread of the fruitfly, led to the discovery that those found in the South had evolved by being selected by new conditions.

Birch was remembered by his students as a teacher of excellence, who promoted opportunities and research. Understanding the changing structural requirements of a university, he combined botany and zoology into a School of Biological Sciences. Birch did much to maintain the academic traditions established at Sydney University by Professors Harrison, Dakin and Murray.

Agriculture and Pastoral Science:
Science for Survival

The use of scientific knowledge and the scientific method to address near constant challenges that threatened the development of cost effective and efficient rural industries in Colonial and early federated Australia became a blueprint for applied science.

For Europeans it was always going to be difficult to grow crops and graze animals in a vast, remote country, with a completely foreign environment. The successful formula was a simple one – identify the problem then solve it using contemporary science, and innovation. The transition of colonial Australia from a gaol of one thousand, to an independent and significant nation of over three million with the highest material standard of living in the world, in little more than a century, is largely due to the success of the rural experiment.

Initially stakeholders formed informal collectives while later colonial governments assisted through departmental structures and agricultural colleges. Once universities were established it became clear that agronomists and veterinarians were needed to translate the body of accumulating science to individuals and their farms and properties. Between 1910 and 1927 departments of agricultural science and veterinary medicine were formed as a second wave of university disciplines, and in a similar fashion to biomedical science they were linked to research institutes. The colonial universities had little interest in teaching technical skills or studying problems on the land. Colonial and state departments did what they could, and where necessary joined with experts to address particular problems such as rust disease in wheat and tick related babesiosis in cattle, on an as needed basis. In 1916 William Hughes, the Prime Minister, recognised the need for a national body that could recruit resources to solve challenges to rural industry. Impressed with models he had seen in Germany where industrial and basic research were better integrated, he established a scientific advisory council to advise and coordinate research activities relevant to primary industries. Five priority areas were identified:

soil surveys, cattle ticks, sheep blowflies, prickly pear infestation and forest products. Notable early successes were the control of prickly pear, the greatest weed problem Australia ever faced, and development of processes for making paper from hardwood pulp.

"The Prickly Pear's Acclimatisation in Australia" in *Commonwealth Prickly Pear Board Report* (1925)

Prickly Pear became a major problem for agriculture and grazing in New South Wales and Queensland. By 1925 it covered 58 million acres, increasing at a rate of one million acres each year. Bio-control of the offending species *Chelinidea tabulata* was through the introduction of the cactoblastis moth, the larvae of which eat the leaves and seedpods. The considerable scientific study required to ensure specificity of damage was funded in part through the first grant given by the CSIR.

Prickly pear was brought to Australia with the First Fleet on the advice of Joseph Banks. He envisaged a red dye industry using cochineal insects, which feed specifically on prickly pear. Nothing came of that, but by the 1830s and 1840s some graziers were using prickly pear as stock food. Spread that way and by birds, the weed advanced through New South Wales and into Queensland. By 1920 60 million acres were being infested each year. Chemical and physical destruction of the weed was ineffective. The new advisory council initiated and coordinated a research programme that led in 1926 to the mass introduction of the cactoblastis caterpillar which consumes and destroys prickly pear.

Around the same time, the lucrative banana industry was brought to a halt by the rapid spread of a destructive disease known as bunchy top. By 1925 ninety per cent of the banana growing area was out of production, with more than 800 plantations deserted. The new council was asked to help. Field studies identified the problem as a viral disease, spread by the banana aphid. Effective control followed a programme of clearance of every affected plant. The dramatic impact of this simple strategy can be judged by an increase from 40,000 bushels of bananas harvested in 1930 to more than three million bushels twenty years later.

In 1926 the council was restructured, put on a better financial base, and named the Council of Scientific and Industrial Research (CSIR). Sir George Julius, inventor of the totalisator, was Chairman and Professor David Rivett the Chief Scientific Officer. In time, support for secondary industries was added to its mandate and during World War II, the organisation's resources were used to develop radar and support other defence projects. Despite funding at levels below those of international competitors, an enormous amount of productive applied research would follow, and the CSIR played a critical role often in conjunction with institutes, universities and government departments.

Wheat

Internationally competitive export of wheat has always been a barometer for success of Australian agriculture. James Ruse in New South Wales was the first successful wheat farmer, but the newer colonies had more success.

Wheat was first exported in quantity in 1845, though exports were sporadic until the 1870s. By then there were around 500 mills across the country and much of the wheat was exported as flour. This would not change until the 1930s, when most client countries had their own mills. By the time of World War II wheat was the most valuable agricultural product, second only to beef in export earnings. Export was controlled through a Wheat Board established in 1939.

Cropping was always defined by rainfall, the wheat belt across south eastern Australia being the temperate buffer between the 12 inch and 20 inch rainfall zones. South Australia was the biggest producer until the 1890s. The limit there to "safe" wheat production was said to be the Goyder line, the 12 inch rainfall isoline, although this limit was challenged by faith and science. The momentum of wheat production moved east, and after 1910 New South Wales became the leading producer. Several factors were important in expansion of the wheat industry, including development of infrastructure, especially railway networks and water delivery, a shift to support farmers by the system of free selection of land, which led to the break up of pastoral leases, and a move towards a "two-legged" economy combining sheep and wheat on the same farm. The contribution of science was a major factor in three main areas: automation of large scale cropping, soil science and breeding of hardy disease-free varietals. At the start, farmers in Australia had to adapt to conditions vastly different to those of Europe. Wheat ripened more quickly but the land could not grow such dense crops. Crop rotation, co-production with sheep and increased acreages calling for wider ploughs and bigger teams were all important.

A crunch came to South Australia in 1843 – an excellent year for wheat. There was simply not enough labour to harvest it. A local competition led John Ridley to develop his stripper which transformed harvesting. Ridley was able to reap and thresh 70 acres in seven days – eventually four men in one day could harvest as much wheat as two could previously harvest in a season!

Innovations in harvesting technology continued and spread. H.V. McKay in Victoria combined harvesting, winnowing and threshing in the Sunshine Harvester. By 1904 his company had become the largest manufacturing

Harvesting at Goonoo Goonoo, a property of the Australian Agricultural Company, in *Thirtieth Report of the Department of Lands* (1909)

exporter and the greatest manufacturing enterprise in Australia. Mechanical inventiveness was able to assist with more than just reaping. One of the most important examples of its achievements for cropping is the ingenious stump jump plough, devised in 1876 by the Smith brothers to ride over obstructions such as stumps and roots so common in mallee country being brought into cultivation for wheat.

In the 1880s yields fell mysteriously to less than 5 bushels per acre across much of the South Australian wheat belt. Concerned that the problem was caused by a soil deficiency, the South Australian Government established Roseworthy Agricultural College with an experimental farm to investigate soil deficiencies. In 1887 William Laurie was recruited from Edinburgh to work on the problem and Australian soils were found to be deficient in phosphorus. Laurie's experiments and energy transformed wheat growing. Use of superphosphate boosted production and increased

the value of land under wheat. These foundation studies on soil deficiency were followed by a series of studies on micro-deficiencies beginning at the Waite Institute in Adelaide, that would lead the world in the understanding of elemental deficiency disease in plants and grazing animals. Australia's leached soils had created an environment that facilitated Australia taking a leading role in soil science.

A third challenge came after farmers introduced wheat beyond the edges of the original wheatbelt in areas with low and erratic rainfall. A severe epidemic of the red fungal disease, stem rust, in 1889 initiated an attempt across colonial borders to coordinate rust control. The importance of breeding rust-resistant strains of wheat had been identified by William Farrer, who set up an experimental farm in New South Wales to develop resistant hybrids. Farrer stood out as a world leader in crossbreeding experiments and his Federation varietal was grown in Australia and abroad for forty years.

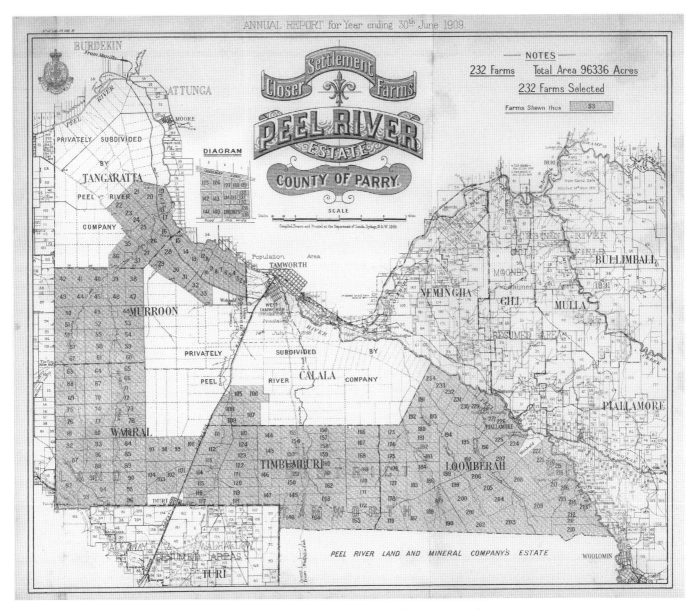

Closer Settlement Farms. Peel River Estate in *Thirtieth Report of the Department of Lands* (1909)

Science, and lack of science, led to a significant shift of the centre of wheat farming from South Australia to New South Wales early in the 20th century. Many factors combined to cause this shift, including Farrer's Federation wheat varietal becoming available in 1902, enabling "safe" agriculture in areas which were dryer and hotter. This put pressure on government to make smaller farming lots available where squatting had inefficiently monopolised available land. Science contributed to an interesting social outcome related to national growth. In 1889 when most wheat was produced in South Australia, New South Wales wheat farms covered 350,000 acres and produced only a quarter as much wheat as South Australia. By 1908 wheat was farmed in New South Wales on over 2 million acres and production exceeded South Australias. Subdivision of the Australian Agricultural Company's land at the Peel River is an example of the New South Wales Government's response of legislating with Selection Acts to break up pastoral holdings for cropping. The Government also doubled efforts to make previously inaccessible land, such as in the Dorrigo region, available for Conditional Purchase Leases.

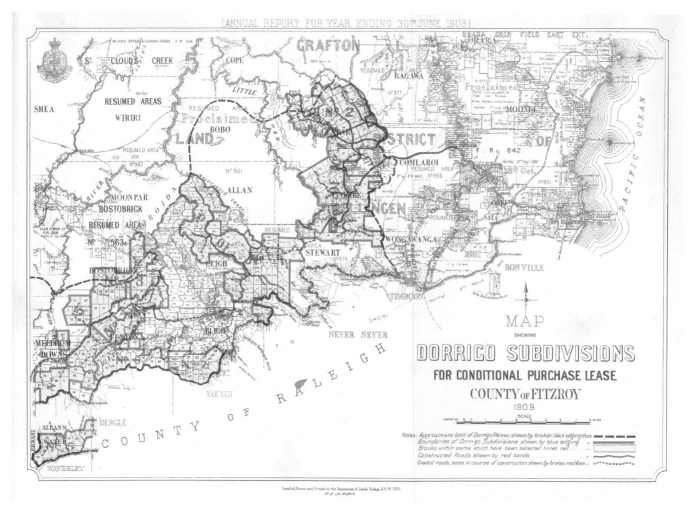

Map Shewing Dorrigo Subdivisions for Conditional Purchase Lease in *Thirtieth Report of the Department of Lands* (1909)

Farrer's breeding strategy would continue in wheat research up until modern times, with an acceptance that all new varietals be rust resistant, preferably based on multiple genetic loci.

Study of the fungal disease that causes rust was pioneered by Daniel McAlpine in the Victorian Department of Agriculture, who became a world leading plant pathologist. McAlpine worked closely with Farrer and would lead the way in understanding rust disease and also bunt and smuts.

The academic scientific achievements of the likes of McAlpine were complemented by the applied science of Alf Hannaford, who came from a farm, and saw the practical benefits of developing machinery that could grade seed for planting and protect it against smuts by pickling the seeds. Hannaford's successful manufacturing began in 1925, and his seed grading and treatment business continues today.

Sheep

The pivotal importance of sheep to the economic growth and prosperity of Australia is reflected in the parallel increase of sheep and human populations. A discrepancy in the 1890s reflected the severe economic recession at the time when many left the country and settled in cities. It has often been said that the wool industry was about "drought, price and rabbits", an assessment which neglects the powerful influence of science on husbandry and wool production, combining a blend of imported and local contributions. As with other rural industries, these early initiatives underpinned subsequent international status of Australian research conducted through institutes, the CSIR, and universities in the 20th century. The sheep industry benefitted greatly from mechanisation, water management, power supply, farm management including fencing, soil and pasture development, and rabbit control.

The capacity to freeze meat for export led to changes in the production balance of wool and meat. Up until 1900 nearly all sheep were Merinos. Then "meat sheep" such as the short wool Ryeland and the long wool Border Leicester became important. They were crossed with merinos to produce breeds that yield both wool and meat, their wool being coarser and stronger than merino wool, with fibre diameters of 32 to 38 microns (fine merino wool has fibres of less than 21). Perfection of the meat chilling process by the CSIR in 1932 led to increases in beef and mutton export to Europe, which by World War II reached about 60 million pounds annually.

This was not the story of the 19th century, when "excess sheep" were boiled down for tallow. The focus was on producing fine wool, and the science was based on clever breeding. This began with John and Elizabeth Macarthur at Elizabeth Farm in 1794 then at Camden Park in 1820, followed by breeders in the Mudgee area and by the Australian Agricultural Company which imported Saxon Merinos, aiming to increase wool density and cover.

The impressive success of breeders and wool sorters in improving merino genetics was reflected in wool production. Macarthur's rams in 1820 produced about 5 pounds of wool each and the average clip was 2.5 pounds per sheep. By century end some rams produced 40 pound of wool, the average clip being around 7 pound. Samuel McCaughey, a wealthy and influential breeder, in 1889 imported very wrinkly sheep from Vermont (USA), aiming to benefit from the increase in the wool bearing surface. Wrinkled sheep quickly became the breeders' preference throughout New South Wales and Queensland. It proved to be a disaster because the new genetically wrinkled stock were less robust in drought conditions and shearing them was difficult. Increases in the weight of the wool clip were deceptive because of the high grease content. The greatest challenge was the susceptibility of these sheep to *Lucilia cuprina*, the only blowfly variety whose larval form invades live flesh – a horrific problem. In the drought years of the 1890s, blowfly strike was as economically damaging as the rabbit plague. Science offered no quick fix, and management was restricted to dipping, crutching and the destructive Mules operation which removes wrinkles surgically.

In 1859 when Thomas Austin had said "the introduction of a few rabbits could do little harm and might provide a touch of home, in addition to a spot of hunting", he clearly did not envisage the fastest spread ever of any mammal. The dramatic proliferation of rabbits would have an untold impact on the environment and economy of Australia. Rabbits competed with fauna and livestock for feed and habitats, and became the principal cause of erosion. The New South Wales Government offered a £25,000 reward for a solution, which encouraged Louis Pasteur's nephew Adrien Loir to come to Australia from the Pasteur Institute in France, to study the problem. A branch of the Pasteur Institute operated on Rodd Island in Sydney Harbour between 1890 and 1898.

A royal commission followed in 1901, but nothing had much success until biological control was introduced. Beginning with studies by the New South Wales Department of Agriculture in 1926, Australian scientists developed myxomatosis as a viral biocontrol of the rabbit plague. In 1934 the CSIR coopted Sir Charles Martin in England to do laboratory tests, which were followed in 1936 by tests by Lionel Bull of the CSIR that demonstrated safety for domestic animals and native fauna. However the programme was discontinued after field studies in arid areas failed to show spread of the myxomatosis virus.

Immediately after World War II the rabbit problem was of such immense proportions that further field trials were performed in the Albury area. On this occasion heavy rains supported an increased mosquito population, giving the virus a vector not present in earlier field studies. Extensive spread made the programme successful. Within two years after the virus was released in 1950, wool and meat production had increased noticeably. Frank Fenner who had worked with Burnet prior to the War on ectromelia (the pox virus in mice) had begun basic studies on myxomatosis predicting resistance in the rabbit population – which of course would happen! He also injected himself with the virus to dispel fears of harm to human beings.

Two other vastly different challenges in the wool

industry would benefit from scientific discovery. Mechanical shears and disease control both led to an explosion in wool production when traditional methods could not keep pace with export opportunities. Fred Wolseley arriving from England in 1854 to begin sheep farming, immediately understood the shearing problem. He developed a series of working models of mechanical shears. Then working with an engineer, John Howard, he produced a marketable product in 1885. In the first demonstration in the Melbourne wool store of Goldsborough Mort, mechanised shearing was no faster than the control hand shearers, but the hand shearers left on average three quarters of a pound more on the sheep. The first field trial was in 1888 on the Dunlop Station near Bourke in New South Wales. Despite an initial reluctance, shearers were soon shearing 120 sheep per day and 184,000 were shorn in two months. By 1900 nearly all sheds used mechanical shears.

Infectious disease of sheep was always a threat. In the eastern colonies, Cumberland disease was having a major impact. By experimenting on the transmission of the infection, Adrien Loir was able to identify it as anthrax. Loir's work in Australia was immensely important, connecting Australia with European biomedical science while establishing a platform with his vaccine studies for future infection and immunology research programmes. Following Federation, three major infectious diseases threatened sheep economy: footrot, black disease and liver fluke infestation. Australian scientists worked out both the pathogenesis and prevention of two of the infections.

Footrot had been endemic in Britain since the 1700s. It was introduced into Australia with European sheep in 1802. It was recognised as contagious and known to reduce wool growth, body weight and lambing. These effects were exacerbated by limits the disease imposed on sheep movement. South Australia and Tasmania were the first to introduce eradication programmes, and Tasmania declared successful eradication in 1939. The breakthrough in understanding footrot came in 1932 when the CSIR began a programme led by Daniel Murnane. He confirmed the infectious nature of the disease in cattle in the ligamentous structures of the foot. For cattle it was a less severe cause of lameness than

for sheep. The same group working in the McMaster Laboratory in Sydney between 1931 and 1939 would isolate the aetiological microbe. William Beveridge used Koch's postulates to identify a gram-negative anaerobic bacterium that he called Dichelobacter nodosus as the single cause of damage of the epidermal tissues of the hoof. He developed a crude vaccine, which was of limited value. In the same group, T.S. Gregory discovered a separate bacterium that caused acute abscesses. By showing that the anaerobic bacteria responsible could only survive a couple of days in the field his studies led to valuable improvements in flock management. The McMaster group in Sydney went on to achieve further improvements in understanding the pathogenesis of the disease by discovering of two proteases (enzymes that can break down tissues) produced by the bacteria that caused hoof detachment, leading to additional management strategies.

Prior to 1930 the most serious infection of sheep was infectious necrotic hepatitis or black disease. It was estimated to cost the industry £3 million annually. This was a golden period for the young CSIR as it focused on economically important infections in livestock. The black disease group was led by Arthur Turner whose careful analysis of the natural history of the disease between 1928 and 1930 first identified the bacterial pathogen, then developed a vaccine which reduced mortality by ninety eight per cent.

Beef Cattle

From an inauspicious and struggling start the beef cattle industry would become one of Australia's major agricultural industries in the 20th century. By the end of this century one quarter of all farmers had beef as a major activity in a $6 billion industry contributing four per cent of Australia's export earnings. Foremost amongst the reasons for this success was the contribution of science to the resolution of three challenges: first managing conditions for herd development including selection and breeding, stock movement to avoid drought, and access to artesian water; second, prevention and control of infectious disease; and third, the development of refrigeration in order to reach the international market.

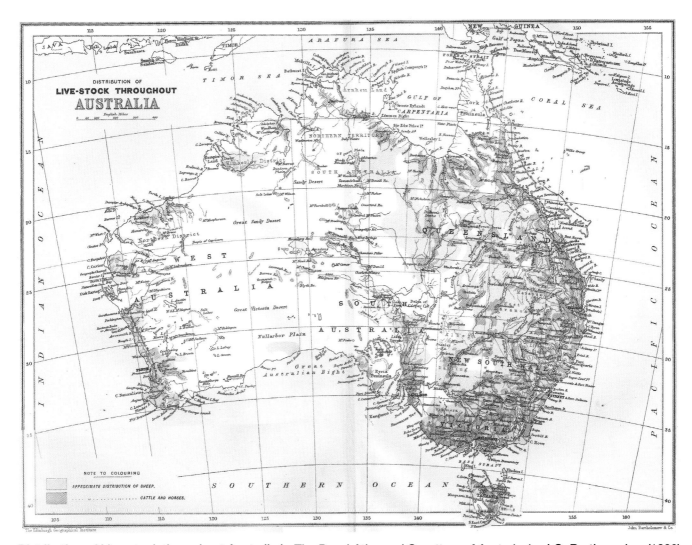

Distribution of Live-stock throughout Australia in *The Royal Atlas and Gazetteer of Australasia*, **J.G. Barthomolew (1890)**

This gazetteer is one the earliest published comprehensive databases of Australian statistics and economic strengths. The thematic map illustrates an extensive distribution of sheep and cattle concentrated along the eastern side of the country, made possible by scientific breeding, control of disease and good stock management.

The role of the man on the land as scientist has been poorly recognised; in many ways the foundation stone to progress resulting in the Australia of today is the growth of knowledge by observation, the very essence of science. While this concept was important in every walk of Australian life in the first 150 years of European habitation, the beef cattle industry provides a classical example. Experimental introduction of different breeds of cattle and subsequent breeding to identify the best environmental fit began while Charles Darwin was recording characteristics of adaptation in species to support his ideas on natural selection. By the end of the 19th century, the leading breeds were shorthorns inland in the west, north and north east, and herefords in coastal regions. Active breeding experiments aimed at continually improving fertility, growth and environmental adaption to the heat and ticks (especially in the north). In the early 1930s R.B. Kelly of the CSIR introduced brahman cattle – a zebu breed originating in Southeast Asia tolerant to conditions found in northern Australia. Kelly's subsequent field studies not only established zebu cattle as a major success in Australian conditions but also began recognition of the importance of scientific studies outside urban centres. His work led to adoption of the santagertrudis (a cross between the brahman, and other zebu breeds, and English shorthorn) into much of Australia. He was followed by a series of geneticists including Helen Turner and eventually (in 1959) the Animal Genetics Division of the CSIRO.

The man who was primarily responsible for the successful development of the cattle industry in the low rainfall country of northern central Australia would be seen as a most unlikely scientist. Sidney Kidman (1857–1935), starting with nothing other than a packhorse at the age of thirteen, would develop a pastoral empire of 60,000 square miles stretching across the centre of Australia. From his observation of Australian conditions, he deduced that by acquiring a chain of stations linking north and south Australia he would ensure pasture and water for his stock at times of drought. In a similar vein, Patrick Durack (1834–98) established linked holdings between south-west Queensland and the Kimberley region of Western Australia, droving over 7,000 head of breeding cattle nearly 5,000 kilometres from his Thylungra Station in Queensland, to the Ord River. It was extraordinary men like Kidman and Durack who had the vision and energy to use land across large segments of Australia, adapting to an ever changing and often hostile environment, to create a business of international economic value to Australia.

The second challenge was infection. Of the three "great cattle plagues" – foot and mouth disease, rinderpest and pleuropneumonia – which devastated cattle holdings in other parts of the world – only pleuropneumonia became established in Australia. It was imported into Victoria in 1858, quickly spreading along the trade route north to New South Wales and Queensland, reaching the Gulf of Carpentaria by 1864, to become endemic in northern Australian herds. One third of the cattle population in New South Wales was lost in two years. The cough of an infected cow could infect others a hundred metres away. In the 1860s crude attempts at immunisation, using fluid from the chests of animals dying of the disease, had little benefit. It must be noted that these "crude attempts" preceded Pasteur and had only Jenner's work for guidance and control as a model. In the 1880s attempts to create a vaccine in the Pasteur Laboratory in Sydney was equally unsuccessful. It was not until Arthur Turner established a laboratory in Townsville for CSIR in 1931 that a successful vaccine was developed. It had an immediate impact on the economy, improving the health of herds, and enabling

Freezing Plant. Ross River Meatworks in Queensland.
Beef Cattle Industry (1913)

Within twelve months of construction of the Ross River Meatworks in 1891, 600 tons of frozen beef had been shipped from Townsville to London, taking advantage of the pioneering work of Eugene Nicolle and Thomas Mort. Meat became Australia's largest rural export, thanks to continued developments in freezing and chilling. The CSIR under the direction of James Vickery from 1933 developed chilling under carbon dioxide, and later, flexible packaging. Vickery revolutionised food science and technology.

movement between states without fear of infection. For the vast areas of the North and Centre, beyond the reach of any vaccination programme, Lionel Rose established a protected area of 3,200 sq km around Alice Springs, to enable sorting of animals from high and low risk groups determined by their area of origin. He used a serological assay developed by Turner to refine classification. These efforts led to a progressive reduction in disease areas, though eradication was not complete until 1973.

Another infection threatening the cattle industry was due to malaria-like protozoa. The pathogens were two related protozoic species, *Babesia bovus* and *B. bigemina*, parasites transmitted by the cattle tick. In 1829 ticks were introduced from Indonesia into what is now the Northern Territory. Deaths were first recorded in 1881 and the disease reached Brisbane and northern New South Wales by 1900. The cause was identified in 1895 by a government pathologist, Charles Pound. The disease was controlled to an extent by quarantine and chemical dips, but three million cattle died from infection and tick infestation in the latter part of the 19th century. It was not until 1964 that an effective trivalent attenuated vaccine, grown in splenectomised calves, became available following successful field trials.

Cumberland disease (anthrax) which troubled sheep was also a threat to cattle, especially in New South Wales. It was first diagnosed in 1847, and initially controlled by quarantine and burning of carcasses. After the building in Sydney of an outreach of the Pasteur Institute in 1890, this experience, and the introduction by Pasteur's nephew Adrien Loir, of the new science of host-parasite relationship study – involving specificity of cause and response – were of immense importance to future research directions in biomedical science.

The third challenge was to transport meat surplus to local need, to Europe. The Industrial Revolution had created a need especially in England for imported meat. Initially live cattle and sheep provided for this need, then various meat preparations. Henry Dangar – known as a surveyor and employee of the Australian Agricultural Company – had raised cattle and sheep and was concerned at the wasteful disposal of excess farm animals by boiling them down for tallow. In 1847 he built a successful canning business to export to England. However the key to large scale meat export was a process to deliver frozen or chilled meat. It became a competitive area. The pioneers were Thomas Mort and a refrigeration engineer, Eugene Nicolle, whom Mort financed and inspired to develop a process. The first successful export was in the *SS Stratheden* which in 1889 delivered a load of meat using compressed air as the refrigerant. By 1889 three Orient Line ships were fitted with Nicholle's freezing equipment.

Water

The greatest challenge to making Australia's remote and hostile regions agriculturally productive was the finding and provision of reliable water supplies. The political imperative of supplying water to rural industry is as critical and unresolved today as at any time in the past. It has reflected the margins of production since Australia was colonised. The importance of rural industry is reflected in Australia's gross domestic product, which in Colonial times and again early in the 20th century was fifty per cent greater per capita than that of the USA, with rural industries contributing about one third of the total (while mining contributed 7–8%). It was the use of scientific methods to protect, maximise and expand cropping and grazing that enabled remarkable economic outcomes, and the creation of an independent nation. In this mix, water was crucial. Regional developments and the introduction of specific irrigation crops were founded on science-based programmes of water retention, access and distribution, involving surveyors, geologists, engineers, government departments and research institutions. As in so many fields that manifest the outcomes of initiative and invention, there is often an inspired and talented individual at the source of each discovery and development.

Three main water conservation and distribution approaches developed: The discovery and use of artesian basins; the optimisation of river water (through strategically placed dams and tanks, irrigation schemes, gravity feed and overflow lakes); and deviation of river flows to avoid waste.

Exploitation of subterranean aquifers opened arid and otherwise unusable land, especially for grazing. The Great Artesian Basin, centred in western Queensland, is supplied by run off from the western slopes of the Great Divide. It covered an area in Queensland and beyond of 1,700,000 sq km – the largest and deepest artesian basin in the world. Henry Russell, the New South Wales astronomer and meteorologist suggested its existence in an article in the journal of the Royal Society of New South Wales published in 1879. He calculated that the Darling River received less of the water falling in its watershed than other rivers. He compared the Murray River which received twenty five per cent while the Darling received only fifteen per cent, so he was able to deduce that large volumes of water descend into subterranean rock strata.

From an extensive knowledge of Queensland geology, Robert Logan Jack recognised the basin like geological formation of western Queensland, predicting the discovery of large aquifers. The first productive bore was sunk near Bourke in New South Wales in 1878, and within a few years 3,000 bores and wells were dug to create valuable pasture in low rainfall areas.

The west flowing Murray-Darling River system was an obvious major water source for inland farms. However competing priorities for the limited supply and the political sensitivities of Victoria, South Australia and New South Wales complicated plans to use the water and have done for nearly two hundred years. Conflict arose in Colonial times between those wanting water to irrigate, those wanting to navigate an internal river highway and those at the bottom end of a compromised supply. Four different types of system were installed over an eighty year period on the Murray and its tributaries.

The first involved a storage and distribution network. An ambitious example was the Wimmera-Mallee project inspired and constructed by John Derry in Victoria in the 1880s. Water flowing off the Victorian Alps and the Grampian mountains was collected from the Goulburn and Loddon rivers and channelled through 10,600 km of pipe via 16,000 tanks to supply agriculture and pastoral land one eighth the size of Victoria, and fifteen towns.

The second system was irrigation directly from rivers, involving complex pumping stations. The first successful venture converted a failed pastoral lease on the Murray River into the "Mildura Colony". It was created by George and William Chaffey, Canadian brothers who had developed a model irrigation system in California. A series of steam pumps moved water from the river into a storage billabong, then by pipes and drains to irrigate 33,000 acres. This visionary concept created a new industry based on fruit and grapes. A similar scheme at Renmark, downstream from Mildura, grew mainly oranges and grapes. Again, it was the Chaffey brothers, but this time in South Australia, without what they described as the "confused help" of the Victorian Government.

A third system took advantage of gravitation between an upstream dam and a downstream weir. One such venture was the Goulburn Valley Irrigation Scheme, which delivered water to a vast, formerly uncultivated area of land in northern Victoria around Shepparton and the Waronga basin for fruit, crops and pasture.

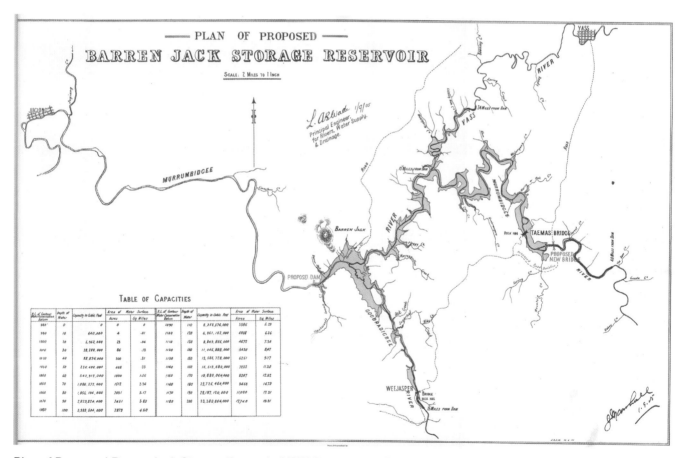

Plan of Proposed Barren Jack Storage Resevoir, *NSW Government Printer* (1905)

In the late 19th century, Samuel McCaughey anticipated an extensive irrigation system based on water retention and supply from the Murrumbidgee River. The Burrinjuck Dam was constructed in 1909 as the fourth largest dam in the world, holding twice as much water as Sydney Harbour, with a catchment area larger than that of the Snowy Mountains. On the back of this system, towns such as Leeton and Griffith were built. The plan was based on a similar system in the Punjab in India.

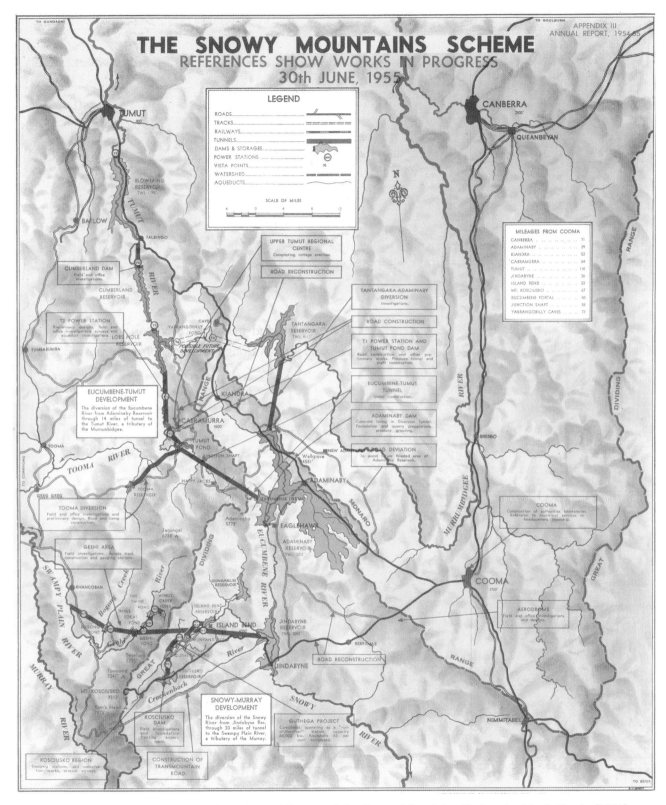

"The Snowy Mountains Scheme" in *Sixth Annual Report of the Snowy Mountains Hydro-electric Authority* (1955)

For over sixty years diversion of the east flowing Snowy River into the Murray-Murrumbidgee River systems had been promoted. The map of Lake George (over the page) was compiled to illustrate an original idea for piping water across the Divide into the Murrumbidgee system — thwarted because Lake George was an unreliable source. The second map records construction of the final design in 1949, that would be completed in 1974 with seven power stations, sixteen dams, 145 km of tunnels and 80 km of aquaducts — at a cost of $820 million!

The Eildon Dam was built between 1915 and 1929, 130 miles upstream from the Goulburn Weir.

A similar development in New South Wales was the vision of a wealthy agriculturalist, Samuel McCaughey. He initiated the Murrumbidgee Irrigation Scheme that would transform unusable land in the region around today's Leeton and Griffith. The centrepiece was a dam on the Murrumbidgee at Barren Jack Mountain (now known as the Burrinjuck Dam) built between 1907 and 1912, with a catchment of 13,000 sq km.

By 1912 2,000 farmers were growing grapes and fruit in the Leeton/Griffith area. A second New South Wales scheme was developed with the Wyangala Dam on the Lachlan River, and a weir at Jemalong.

To measure the amount of water used by each irrigator Jon Dethridge in 1910 invented a clever device consisting of a drum with vanes known as the Dethridge meter. This device was adopted internationally.

A fourth system, aimed at stabilising water flow in the Murray River, arose from the politically sensitive dilemma of appeasing navigators and irrigators at different sites along the river. Unintended outcomes of increasing demands on river flows included disruption of the water table and salination, which still challenge science for solutions. A meeting of stakeholders in 1914 led to construction of the Hume Dam east of Albury, with a capacity of over three million megalitres covering 13,000 hectares, constructed between 1949 and 1960. Controlled release from the dam mitigates the sharp seasonal fluctuations in flow from the upper Murray.

Elsewhere, where lands near river systems had high water tables and low salt content irrigation has presented fewer challenges. Advantage was taken of this circumstance in Queensland, where the Lockyer Valley was made available for vegetables, and the Burdekin River delta provided 12,000 hectares for sugar cane.

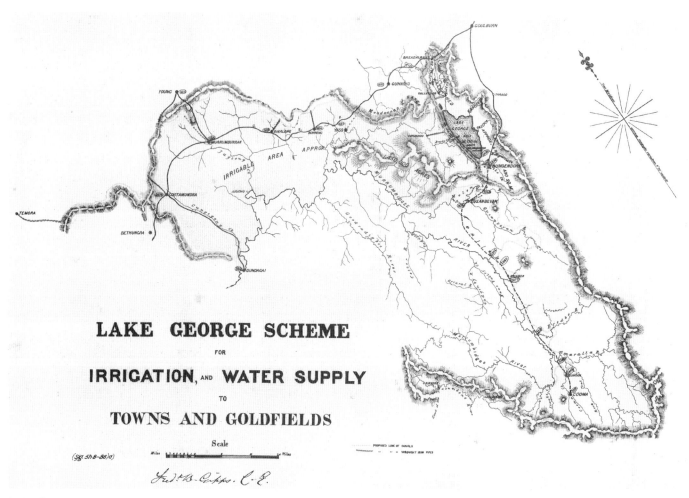

Lake George from Various Surveys 1832–71, 77, J. Donkin

Finally came the approach of diverting river flows: a complete redirection to the west of water flowing eastward off the Great Divide, through a vast array of pipes and dams. It is a monument to the science and skills of engineering. It was first considered in an 1884 proposal to divert water from the Snowy River and collect it in Lake George north of Canberra then pump it across, the adjoining range at its lowest at Slack's Creek. The idea was resurrected in 1938: "I am fully cognisant of the magnitude of the scheme, but great national necessities warrant ponderingly bold measures of relief." (*Canberra Times*, 22.10.38). Promoting the sciences, J. Donkin used an 1880s map of Lake George created by the Government Astronomer Russell, whose study of seiches (stationary waves in a closed system) was thought to explain the rapid variations in the water level of Lake George which led to the initial proposal being dropped. (More recent studies show that the variations can be explained by differences in seasonal inflow and evaporation). The eighty year old idea led to a new plan implemented after World War II, one of the greatest engineering and science driven projects in the world at the time. Construction took twenty five years and involved over 100,000 people from more than thirty countries who added a social mix to the workforce contributing significantly to the intellectual growth of the country. Sixteen major dams, seven power stations, a pumping station, 145 km of trans-mountain tunnels and 80 km of aqueducts combined to make the Snowy Mountains Scheme a civil engineering wonder of the modern world.

Water management began in colonial Australia with tanks on the Tank Stream flowing into Sydney Harbour. Barely 150 years later, driven by scientific method, it encompassed a massive rearrangement of the drainage of the Australian Alps.

Important Scientists in Agricultural and Pastoral Science

John Ridley (1806–87)

John Ridley was a case study supporting Immanuel Kant's idea that the essence of the Enlightenment was to dare to use your own intelligence. Essentially self taught, with a capacity to see beyond the square, he brought his young family from England to South Australia in 1839 for the opportunities Australia offered. Ridley was an entrepreneur, taking advantage of the drought and depression of the 1840s to make clever land and copper mining investments, while developing a flour milling business. He was only in Australia for fourteen years, but in that time, he revolutionised the wheat industry that was threatened by the inefficiency of hand reaping.

South Australian farmer, John Bull, had been experimenting with a mechanical harvester, and had presented it to the Adelaide Corn Exchange in 1843 where a prize was offered for an innovative harvester. A reaping machine invented in Scotland in 1828 worked on the principle of the lucerne mower, enabling "long stalk wheat" to be bound into sheaths for later threshing, leaving straw as a by-product for indoor winter housing of domestic animals – not very useful in Australian conditions. Bull had discovered the principal of a horizontal projecting comb with revolving beaters driven by a belt from the carriage wheels. He didn't win the prize but Ridley, who had been improving Bull's principal, produced a practical stripper that successfully reaped 70 acres of wheat in a week at a cost of 5 shillings per acre. He planned and manufactured his machine – within five years there were fifty machines working locally, while others were exported. With all his investments, Ridley became wealthy and returned to England in 1853. Australian initiative continued to revolutionise agriculture in Australian conditions, with the design of a stripper-harvester by Hugh McKay in 1884 – this separated the grain from the chaff, leading to his development of the Sunshine Harvester machine. His extraordinarily successful commercial venture continued producing agricultural machinery into the 1980s. These three practical farming men saw a problem, each contributing to an efficient answer to the overwhelming volume of wheat in conditions of growth and circumstances of need. They transformed the commercial face of wheat farming in Australia, using scientific principals of observation and experimentation.

William Farrer (1845–1906)

William Farrer arrived in 1882 with two important traditions – he was raised on a farm and had graduated from Cambridge with a background that included science and mathematics.

He had some Classical learning and experience in surveying. Tuberculosis terminated his medical studies, and he migrated to New South Wales for the air, without plans for the future. A distant relative of Charles Darwin, he would later acknowledge Darwin's influence on his ideas on selection and on his observations with respect to blight resistant apples in the USA, phylloxera resistant grapes, and even sugar levels in beet.

Farrer arrived unknown in the Australian community, at a time of crisis in the wheat industry which was performing a poor second to wool and focussing the animosity of graziers whose former leaseholds were being handed over to selectors for cropping. The viability of wheat was challenged by harsh environmental conditions and rust infection.

Farrer saw a way forward by breeding using selection along the lines described by Darwin in *Domestication* published in 1868. When he published his aim of breeding a rust resistant strain of wheat in a series of articles in *The Queenslander* (1882–83), he was condemned by some as a "nobody". He drew further scorn when he differed from experts on the type of rust infection that was of most concern. Faced with detractors, Farrer never moved from his premise that liability to rust was "not mechanical but constitutional" and his paramount belief in an approach to wheat problems based on heritability, variation in rust resistance, and the power of selection.

Darwin's ideas were not readily accepted and Australian experts responding to his proposals in letters to *The Australasian* in 1882 were both arrogant and ill informed. Farrer responded with further principles he would adopt: co-selection of varieties with good milling properties and artificial crossing with naturally selected rust proof varieties.

When he established a small, self funded experimental farm on the Murrumbidgee River (between 1886 and 1898) the sheer importance of his project brought him an appointment from the New South Wales Government. He obtained wheat varieties from around the world but concentrated on some obtained from Canada and India. An Indian varietal he selected matured early which had the dual advantages of protecting growing wheat from hot dry summers and shortening the time during which it was at risk of contracting rust. Late maturing Canadian Fife variety had excellent milling and baking qualities. Farrer was the first breeder in Australia to pursue quality and quantity of wheat together, and

"Emptying a Stripper" in *Wheat in New South Wales* (1908).

A booklet prepared for the 1908 Franco-British exhibition includes a summary of contemporary science-driven farming practice, and shows a "shift to the east" in wheat production (from South Australia to New South Wales). Paradoxically, this stripper was invented in South Australia, and began the mechanisation of cereal production as a key element to economic success.

William Farrer from the $100 note (1984)

worked with a chemist, F.B. Guthrie, to assay hybrids for their baking quality. The first commercial variety generated by selection methods was Bobs, derived from Canadian/Indian hybrids. Yet despite the promising qualities of Bobs, and the Yandilla and Comeback varietals bred from it, farmers persisted with old varietals, such as Purple Straw which was more amenable to Australian harvesting and gave a good yield in good years. Farrer then hybridised Yandilla with a Purple Straw varietal, which after further selection, gave rise to Federation, Farrer's signature wheat. Federation became dominant in Australia between 1910 and 1925, and was widely grown internationally for forty years. It combined good yield with short straw (excellent for Australian processing conditions), and rapid maturation to evade rust. By 1914, of the twenty nine recommended species in New South Wales, twenty two were Farrer hybrids. They enabled wheat to be grown in drier and more rust prone areas, and were largely responsible for the four-fold increase in Australian wheat production between 1897 and 1915.

Farrer's crossing experiments led the world at that time in wheat research. He combined an awareness of Australian conditions with a global view. He considered the whole of the chain from grower to consumer, a concept that had not previously been explored. A practical man who published little, he was at the same time intellectually honest and generous, communicating with international experts, providing them with seed. The breeding methods he pioneered and the stimulus he gave to research in Australia and internationally led to the development of the new and better wheats that superseded Federation and other Farrer wheats. Nevertheless Farrer's Florence was still the second most common varietal at the time of World War II, and even the highly successful rust proof Gabo of the 1950s was influenced by Farrer's work.

Farrer's work on plant genetics went beyond Darwin's understanding of artificial selection. Before Mendel's genetics experiments became known amongst the scientific community, Farrer grasped the heritable nature of disease resistance, maturity and grain quality as well as demonstrating independent segregation of inherited properties within second and subsequent generations after crossing. He summarised much of his scientific work in a paper published in *The Agricultural Gazette of New South Wales* in 1898. Perhaps the best summary of Farrer the scientist is Darwin's description of the ideal natural historian: "Indominable patience, the finest powers of discrimination and sound judgement must be exercised over many years. A clearly predetermined objective must be kept steadily in view. Few men are endowed with these qualities."

Daniel McAlpine (1849–1932)

Daniel McAlpine was a towering figure in plant pathology; a man of his times arriving in Australia as the wheat industry was under enormous pressure to adapt to a hostile environment and the pathogens that thrived in it. Although he was identified as one of the two best English plant pathologists of his time, it is uncertain where he obtained his training and when interest began. Certainly he communicated and exchanged specimens with Mordecai Cooke, of the Royal Botanical Gardens at Kew, who had written the standard text on Australian plant fungi. Although he had no degrees McAlpine came to Australia anticipating a teaching position, supported by connections to Thomas Huxley (a biologist), Archibald Geike (a geologist) and William Thiselton-Dyer (Director, Kew Botanical Gardens), and a record of teaching natural history and introducing illustrated teaching aids for biology students. Like William Clarke and William Farrer before him, McAlpine represents the British immigrant with an enthusiasm for science, who came to Australia for a better life. Pursuing economic goals, career opportunities, or improved health, such men were very different to those commissioned by Banks in the first thirty years of the colony, or those next who explored and asked questions, sending specimens back to England. This third wave of scientists came to settle in Australia and call it home. Driven by enthusiasm and ideas they worked outside the institutional mainstream of science stimulated by need based opportunity and working alone or in government departments. Their contribution through science to the creation of modern Australia was enormous. McAlpine was a signature member of this group – even to the extent of being unrecognised except by a few of his scientific peers. That group included Farrer, Clarke, Ashurbton Thompson and, home grown scientists such as Tebbutt and Hargreaves. They were the life blood of science in colonial Australia, contributing through local Royal Societies and the scientific and lay press.

The rust epidemic that destroyed the wheat crop of 1889 was the defining event of McAlpine's future professional life. The Victorian Department of Agriculture responded to the epidemic by creating the position of "vegetable scientist" with the broadest of job descriptions. But the grand description and high expectations were not matched by local recognition or support. For many years McAlpine worked from home with his own equipment and he had to continue his lecturing positions for financial support: even the University's Professor of Botany ignored him. Yet he was remarkably productive, with authorship of six books beginning with *Systemic Arrangement of Fungi* (1895) – an enormous task that included data from 310 papers, many of which were his own. The previous reference, Cooke's *Handbook of Australian Fungi* included seventy two species of rust – McAlpine included 161 species. Life cycles and pathogenesis of rust disease were described in detail. It brought him international attention.

His *Rusts of Australia*, and *Smuts of Australia* which followed in 1910, were groundbreaking, documenting the two most devastating wheat infections that threatened the wheat industry. Committed to Darwinian ideas – perhaps reflecting Huxley's influence – McAlpine believed the only solution to the problem of rust infection was to cultivate rust resistant wheat – to choose early maturing cultivars and to sow early. He welcomed the opportunity to test rust resistance in collaboration with Farrer who achieved early maturation by creating genetic hybrids.

McAlpine demonstrated that smuts is a more complex infection. He identified transmission by release of spores from flowering plants to invade healthy seed, from threshing, or by deposition in the soil. He developed methods for treating each form of transmission, including application of boiling water and of fungicide and crop rotation.

Bitter pit was a catastrophic disease of apples destroying in some seasons more than half the national crop. Apples were of substantial economic importance at the time in south east Australia. Control of bitter pit became the first project coordinated by the new CSIR committee; McAlpine as the leading plant pathologist in the country was seconded to lead the investigation. McAlpine accepted the job for the "good of Australia," realising that bitter pit was not an effect of infection or toxins, but a disturbance in function that would not easily be identified. In fact it would be over thirty years before the physical and micro-nutrient aetiology was fully understood. Meanwhile McAlpine's practical suggestions were welcomed by farmers. However his fundamental discoveries were not recognised by government or Melbourne University's Department of Botany.

James Prescott (1890–1987) and the scientists of the Waite Insitute

English born and internationally experienced, Prescott became a world authority on the science of Australian soil from a chemical and geophysical viewpoint. He was recruited in 1924 to the chair of agricultural chemistry at South Australia's Waite Agricultural Research Institute after completing studies on phosphate deficiency at Rothamsted Experimental Station in England, and on nitrogen supplementation in Egypt. His first studies at Waite were to understand Australian soils and the impact of leached laterite and variations across the continent. He began working with the CSIR through David Rivett and in 1931 published the first map of soil zones noting the impact of climate and vegetation.

He studied wide ranging soil issues including salinity in irrigation areas in relation to shifts in the water table. He developed a soil/water index for regions as a guide to the length of the growing season and climate limits to agriculture, by combining historic rainfall and temperature records with air saturation estimates. He was widely recognised for his contributions to understanding and managing conditions far different to those in Europe where glaciation had refreshed the soil.

Prescott noted at an International Symposium at the Waite Institute in 1975 that "no country has deserved greater benefit from trace elements than Australia". It was not by chance that Australia led the world in soil nutrient research as much of its soil was old, weathered and leached. Early agriculturalists were confused by variations in crop growth and seemingly inconsistent effects of different soils on different crops.

We now know that there was only limited Pleistocene glaciation in Australia (compared to Europe and New Zealand where newly formed soils left by glaciation had been subjected to no more than ten thousand years of weathering). In Australia there were large areas of old lateritic soil or more recent soils formed following many cycles of erosion, leaching and deposition. Early studies by William Lawrie at the Roseworthy College near Adelaide showed deficiencies of phosphorus and nitrogen especially in arid areas of western and central Australia. These deficiencies were reversed by massive superphosphate applications and use of legumes in crop rotation for nitrogen fixation. Later, potassium and sulphur were found to be deficient in some areas. This experiment of nature, creating old leached soils, would catapult Australia into international leadership in discovery of trace nutrient deficiency in agriculture. The patchiness of impact of such deficiencies was confusing. It became clear that plants generally shared similar requirements, but differed in their capacity to acquire, retain or recycle nutrients in any particular ecosystem. Native vegetation on nutrient-poor soil had evolved more efficient systems. Monocot plants (especially grasses) have less pectin in their cell walls to bind calcium and boron than dicot plants, and thus have a much lower requirement for those minerals.

In 1924, when he donated the land on which to establish the Waite Institute, Peter Waite wrote of the success of South Australian innovators. "With comparatively little training they have placed our wheat, wool and fruit in the highest estimation of the world, our agricultural machinery is good enough for the Americans to copy, and our farming methods . . . are accepted in other states." It was time "to call science to our aid to [an even] greater extent". Waite was not disappointed, as the Institute (often together with the CSIR and the South Australia Department of Agriculture) would become recognised as Australia's foremost agricultural research institute, and under James Prescott a global leader and international model for agricultural research.

Work in South Australia led by William Lawrie at Roseworthy College in the latter part of the previous century with a focus on macronutrients, had inspired Peter Waite. Now attention turned to micronutrient research. By the time the Waite opened, the big picture of macronutrients, namely nitrogen, phosphorus, sulphur, potassium, calcium and magnesium, was reasonably understood. Yet South Australia was losing ground in agriculture to other states, and variations in soil productivity often specific for one or another nutrient, led to reduced yield and quality of particular crops. Over thirty years from 1928, the Waite led the world in research into micronutrient deficiency, implementing diagnostic and remedial programmes, for deficiencies of soil in manganese, iron, zinc, boron and molybdenum. This period began with a demonstration that application of magnesium salts increased crop growth on the Yorke Peninsula twenty-fold!

The first studies showing manganesem deficiency in 1928 were led by Professor C.S. Piper, a chemist who developed a method known as solution culture, which he continued to use in conjunction with field studies such as those conducted by Professor David Riceman from the CSIR on the Limestone Coast at Robe. Riceman and his team conducted many world first field studies for cobalt, zinc and copper deficiency, an achievement which was later widely recognised as "Desert Conquest". Important novel contributions included Alf Anderson's field study showing molybdenum promoting subclover pasture growth in 1942 in the Adelaide Hills, and then in 1946, promotion of pasture growth in south east South Australia using copper and zinc.

Such was the range of problems and deficiencies, that a chequerboard approach could be used to identify specific regional deficiencies. In 1925 the Permanent Rotation Trial was established on thirty four adjoining plots, seven remaining in unbroken sequence for over seventy five years, the world's longest trial sequence. This enabled discovery that fallow phases were associated with the greatest fall in organic carbon, unlike pasture rotations, which made little difference. The findings informed important discussions on carbon retention to maintain soil structure and nutrient availability.

In 1935 coast disease – a wasting disease of sheep feeding on calcareous coastal pasture – was found on Kangaroo Island to respond dramatically to cobalt. Identification of cobalt deficiency was another world first, this by Professor E. Lines. Subsequent studies found co-deficiency with copper was linked to acute coast disease, and that cobalt-deficient sheep suffer a reduced production of vitamin B12 by the rumen microbiota. CSIRO studies showed that staggers in sheep could be prevented with cobalt which protected against neurotoxins from particular crops. Thus, the work initiated in South Australia as a result of an industry challenge related to soil deficiency led to studies of complex relationships between pasture and grazing animals and basic physiological questions related to the intestinal microbiome.

Through his time as Professor of Agricultural Chemistry at the Waite Institute (from 1924), Prescott was the accepted leader of soil science research. He commanded international respect and was a leader of national importance, not just at the Waite, but through his association with David Rivett and the CSIR from 1927, when he combined his laboratory research with field studies, including assessment of salinity in irrigated areas, and ran successful workshops. He continued his leadership role with the CSIR until 1947 and retired from the Waite in 1955.

Robert Logan Jack (1845–1921)

Robert Jack was born and educated in Scotland. He was recruited as geological surveyor for northern Queensland in 1876, after a creditable record contributing to the geological mapping of Scotland.

His contribution to recognising and mapping the geology of Queensland through the latter part of the 19th century, put him in a good place to recognise the massive extent and character of the Great Artesian Basin extending across much of western Queensland. He recognised the aquifer as Triassic, Jurassic and

Cretaceous sandstone, which allowed him to "connect the dots" made by a drilling programme he initiated in 1881. He visualised a large synclined trough connected to the Great Dividing Range in the east in which porous sandstone aquifers collected water from the western slopes. Jack determined that water was confined to the sandstone by impermeable marine claystones, a continuous geological formation he deduced by examining rock sequences from bore holes.

The first artesian bore into the Great Artesian Basin had been drilled at Kallora Station near Bourke in north west New South Wales in 1879. Following a severe drought, the Queensland Government commissioned Jack to determine the potential value of the artesian system. His drilling programme established that vast areas in western Queensland were likely connected to the artesian basin. Commercial drilling began in Blackall, Cunnamulla and Barcaldine, and soon after in a multitude of other locations.

Jack's contribution to understanding the geology of Queensland was more than defining artesian basins. He was involved in mapping the geology around mining sites, especially where gold has been found. His findings were incorporated into government reports, and texts.

Arthur Turner (1900–89)

Arthur Turner is an outstanding example of an Australian 20th century new age scientist. Born in Melbourne of poor parents and publicly educated, he graduated in veterinary science in 1923. He became a microbiologist with the CSIR and established a platform for translational science in Australia by discovering the pathogenesis and treatment of the most destructive infections in sheep and cattle. His studies on host-parasite relationships and disease established a paradigm that continued in Australian science, culminating with Barry Marshall and Robin Warren discovering the role of *Helicobacter pylori* in gastritis and peptic ulcer disease for which they won a Nobel Prize in 2005.

Turner first turned his attention to black disease in 1924, working as a fellow in the University of Melbourne. Black disease or infectious necrotic hepatitis, was a fatal disease of sheep, and less commonly of cattle, pigs and horses, first recognised in Australia in 1894. By

"Map of Australia Showing the Extent of the Known Artesian Basins" (1914) in *Report of the Second Interstate Conference on Artesian Water* (1914)

This map illustrated conference presentations warning of limitations of artesian bore water, having scientifically accessed supply, measured at 40% reduction in flow over ten years.

1930 it had become the most serious infectious disease of sheep, costing the industry about £1 million pounds per year. Turner was awarded a Rockefeller Travelling Fellowship in 1926, and took bacteria isolated from sheep with black disease with him to the Pasteur Institute in Paris. There he identified the pathogen as an anaerobe, *Bacillus oedematians*, and began studies towards making a vaccine. He returned to Melbourne in 1928 after a stay in the Cambridge Institute of Animal Pathology, to join the CSIR under David Rivett, with his laboratory in the

Veterinary Research Institute. There he began a detailed study of the host-parasite relationship in sheep. He demonstrated that damage to the liver by a migrating liver fluke (*Fasciola hepatica*) accessing bile ducts, causes necrosis of tissue and anaerobic conditions favouring germination of resident spores with release of toxin and clinical disease.

In 1930 Turner showed the value of developing a model in small experimental animals, here the guinea pig, defining two lines of disease prevention. First, eradication of the snail (*Lymnaea tomentosa*) as the intermediate host of the liver fluke, by treating water with copper sulphate. Second, he developed an inactivated vaccine from six strains of *B. oedematians* (now called *Clostridium novyi*). After the addition of an adjuvant to Turner's vaccine, to allow protracted immunity from a single dose, it was used internationally to control black disease.

Turner's success in Melbourne led to a five year secondment to establish a research experimental station near Townsville in 1931, to study three major diseases threatening the northern cattle industry, in conjunction with a breeding programme based on zebu cattle introduced from Texas crossed with British stock. These diseases were a respiratory infection caused by pleuropneumonia-like organisms (PPLO), redwater fever (babesiosis), and pegleg (which proved to be a deficiency disease rather than an infection). Turner decided to concentrate on PPLO. Even at that time, he found the bureaucratic obligations of his position onerous. PPLO induced disease was economically the second most serious livestock infection after black disease. It had become endemic in the north by 1926. At the time, the structure and nature of the pathogenic organism was not well understood. Turner made breakthrough discoveries, developing a cell free medium that allowed profuse growth of the microbe, and discovered the means of attenuation of the organism which enabled the production of a stable vaccine known as VF5. He completed important studies on the morphology and function of PPLO, which would not be identified as a bacterium lacking a cell wall until the 1940s. Turner also developed a complement fixing antibody test to enable detection of infected cattle, a leap forward in herd management.

Frederick Wolseley (1837–99)
Herbert Austin (1866–1941)

Frederick Wolseley and Herbert Austin were remarkable men born and raised in Ireland and England respectively, who seeing no future at home, left for opportunities in Australia. Their success in colonial Australia and their return to Britian were followed by pioneering roles in the motor industry. They exemplify the previously unimaginable success in Britain that some people attained after returning from time in an Australian environment, taking opportunities available.

Wolseley was typical of inventors who used scientific method, to create important markets in Australia and abroad. He built up land holdings as a squatter and recognised the limitations of hand shearing that faced an industry with 100 million sheep needing to be shorn. Beginning in 1868 with no mechanical training, over twenty years he developed a series of models of mechanised shears. He held his first public demonstration in the Goldsborough Mort woolstore in Melbourne in 1885. This demonstration showed little difference between mechanical and hand shears in the time taken to shear a sheep. However compared with controls using hand shears, shearers using the mechanical shears were able to recover three quarters of a pound extra wool from each sheep shorn. By 1900 nearly all sheds used mechanical shears. Wolseley established a factory in Melbourne to manufacture mechanical shearing equipment. In 1889 he also set up a Machine Manufacturing Company at Birmingham in England.

In 1884, unable to get training in mechanics at home, Herbert Austin migrated to Australia at the age of eighteen to take an apprenticeship in Melbourne. He met Wolseley who was about to start his Melbourne factory. Austin devised improvements to the shearing machinery and Wolseley then employed him in starting the Birmingham factory and managing the business. Austin reorganised the company, diversifying the product range, and developed a particular interest in manufacturing a motor car – the Wolseley Autocar No. 1. It began the British car industry. Austin solved an early problem in car design of a cumbersome transmission and gear system, by inventing the "universal gate change".

"Artesian Bore: Moree District" in *Thirtieth Report of the Department of Lands* (1909).

Bores tapping the Great Artesian Basin to supply water for livestock and irrigation date from 1878. Careful mapping and identification of the porous rock formation determined the largest artesian basin in the world at 1.7 million sq km.

By 1900 the Wolseley car was the company's main product. In 1905 Austin established his own company, and the first Austin car was sold in 1906.

Wolseley and Austin were innovative and persistent men who had to come to Australia to find the challenges and opportunities that enabled them to revolutionise the wool industry. Their major role in establishing the British motor industry and Herbert Austin's impact on car making worldwide could not have been accomplished without the start that Australian conditions gave their inventive careers.

Lionel Bull (1889–1979)

Lionel Bull, one of the first graduates of the Melbourne veterinary school, pioneered veterinary medical research in Australia, with a remarkable capacity to focus on the problem of the day in the context of contemporary science. His twenty years as head of animal health in the CSIR from 1934 took him through a productive period working on host-parasite relationships in domestic animals, to the emergence of animal genetics and its revolutionary effect on breeding, and the introduction of pharmaceuticals. His studies in the CSIR included tuberculosis in cattle, rabbit control by myxomatosis, production of the V5 vaccine for the prevention of pleuropneumonia, exotoxin production by clostridia, bovine haematuria in Queensland and toxaemic jaundice in sheep. In his early studies Bull identified pathogens including *Cryptococcus neoformans* using India ink to stain their capsules, and the organism that caused "lumpy wool" in sheep. He described the natural history of caseus lymphadenopathy in sheep, showing its transmission through cuts at the time of shearing.

In later studies he examined the interaction between cattle and pyrolizidine alkaloids. Three per cent of plants contain these alkaloids which protect them from insects. Many of them also cause liver damage in domestic animals. This study continued the theme of Bull's work – interaction between animal and external factors including microbes and toxins. As Head of Division and through most of the academic veterinary journals and organisations, his leadership was widely respected. His hands-on style did not suit everyone, and the friction between Bull and H.R. Marston was widely known. Marston was an outstanding biochemist who with colleagues found cobalt deficiency in soil as the cause of coast disease in sheep. It was Marston who discovered that deficiency of trace elements copper and zinc rendered South Australia's Ninety Mile Desert unfit for farming and that it could be reversed by supplemental addition of the missing elements. Marston's role in research and administration was important to him, and being responsible to an autocratic Bull, created problems. Bull's deficiencies in administrative skills can not detract, from the pivotal role he played as a scientist through the formative years of the CSIR at local and national levels. He personified the "whatever it takes" aptitude to solve the challenges facing the breeding and care of sheep and cattle in the first half of the 20th century.

James Harrison (1816–93)
Eugène Nicolle (1823–1909)
Thomas Mort (1816–78)
Edward Hallstrom (1886–1970)

In early colonial times cattle raising and shepherding were often combined. Practical difficulties of unfenced grazing kept livestock numbers down. Greater security of land tenure encouraged fencing and by 1830 with more cattle than needed for work or meat, the average value of cattle plummetted from about £10 per head to 10 shillings. There was no economical way to supply overseas markets with meat from surplus Australian livestock until refrigeration was developed. By 1850 ten per cent of Australia's sheep were boiled down and 11,000 tons of tallow per year were exported. In Sydney Dr Lang lamented the waste of meat "with millions at home [in Britain] on the brink of starvation". Henry Dangar in Newcastle in 1847 began canning meat, which was marginally valuable, though the English never developed a taste for canned meat.

Mechanical refrigeration came about because of three men – James Harrison, Eugène Nicolle and Thomas Mort. Harrison and Nicolle were the inventors. It was Mort who backed belief with funds for the long and expensive experimental process. The main experimental model was set up in the yard of the Royal Hotel in Sydney in 1872 for a fifteen month trial. The experiments cost Mort £80,000, an enormous sum at that time.

James Harrison was a newspaper proprietor whose hobby became refrigeration. At that time, the only way to refrigerate food was with ice imported from North America. Harrison communicated with Faraday and Tyndall, English scientists interested in latent heat and fluid-gas transformation. The system he invented was based on mechanical vapour-compression using ether, passed through cooling coils, in a closed recirculating system. Harrison's invention was tested in a brewing company in Bendigo in 1854. It was driven by a 5 metre flywheel and produced 3,000 kg of ice each day. His initial attempt at using freezing mixes to transport meat to Europe failed and it was never going to happen with an ether system.

Eugène Nicolle, a French trained engineer, arrived in Sydney in 1853, and with his training and experience in engineering had no difficulty in getting employment. His early interest was ice making and in 1861 he formed the Sydney Ice Company with Harrison, who was expanding his interests to Sydney, and Peter Russell whose engineering company would make the machines. Nicolle developed many improvements, pioneering heat exchange systems and enhancing the mechanics of the production. The idea of freezing meat for export was immediately understood by Thomas Mort who had developed a prosperous trade and business in sheep and wool. In 1875, at a famous lunch at his new freezing works at Bowenfels (near Lithgow), where of course frozen meat was served, Mort's enthusiasm and vision were evident: "there is no work on the world's carpet greater" he said, than frozen meat, and he envisaged a utopian world based on food exchange, because "science has drawn aside the veil."

In repeated experiments Nicolle used "high pressure" and "low pressure" ammonia and air expansion, yielding a stream of patents and inventions, and repeated disappointments. On the brink of a successful trial, with a ship modified to transport frozen meat to England, the captain refused to risk leakage of ammonia at sea, noting that insurance would not cover the damage. In the race between Australia, New Zealand and Argentina to be the first exporter of intact frozen meat, Australia's 1879 shipment to England was beaten by an Argentinian cargo that reached France. By 1881 three Orient Line vessels were successfully fitted with freezing equipment designed by Nicolle and the frozen meat export industry was established. By this time, with Mort no longer alive and Nicolle retired, Plant was manufactured in England.

Edward Hallstrom would continue Australia's preoccupation with refrigeration – driven largely by its environment – by extending refrigeration to the domestic market. Largely self educated and from a poor background, Hallstrom had left school at thirteen. He was highly intelligent and motivated, and combined his entrepreneurship with innovation to develop refrigerators following research and experimentation in his Sydney backyard. They were powered by kerosene and sold in country areas. He quickly diversified, and the upright Silent Knight model, run on gas or electricity, became a by-word in Australian society. In the mid-1940s 1,200 refrigerators were produced each week.

Hallstrom lent his skills to like minded inventors, especially those who followed Hargreaves in experimental aviation, and those involved in anaesthesiology. His new wealth enabled him to fund a life long passion in natural history, especially zoology. He became the great promoter of the Sydney Zoo, funding initiatives and introducing effective breeding programmes at a time when supervised breeding was becoming scientifically important to the preservation of species, and initiating scientific expeditions in Papua New Guinea out of which came exchange programmes and projects aimed at preserving the bird of paradise. In later life he clashed with "new scientific zoology", reflecting a time related phenomenon of "old science" versus "new science" and experience versus theory approach to problem solving. His generosity over many years led him to become one of the main private beneficiaries to science, particularly medical science. Hallstrom was a truly great Australian whose success based on science was shared widely and unselfishly.

Anthropology:
A Focal Point

From first contact, Europeans were fascinated by Australian Aborigines. By the end of the 19th century, serious scientists such as Herbert Basedow saw anthropology as a branch of natural history, consistent with the classification of Linnaeus and with Darwin's theory of evolution.

Over two hundred years there was a transition in ideas, from seeing native Australians as a race distinct from and inferior to Europeans, to recognising that a dynamic and diversified Aboriginal population is defining itself rather than fitting into any racial division decreed by historic tradition. There were periods of belief in unilinear social evolution along Darwinian lines, then of cultural and social anthropology involving mechanisms such as cultural diffusion.

The concept of race was fundamental to 19th century anthropology, with man included in Linnaeus' classification system. It was a period for measuring bones and skulls and it came easily to researchers to identify the Australian Aborigine at the lowest level in a classification of kinship types, which Jeeves Morgan later codified. By 1880 Darwin's conception of evolution by natural selection had engulfed all biological science. It had become "the great unifying idea" so long sought by naturalists. Anthropology was simply "social evolution", and racial characterisation was a natural fit. Race remained a fundamental of research well after evolutionary theories had been replaced by social anthropology in the 1920s, led by British studies that became known as structural functionalism.

There were two important collaborative studies of Australian Aborigines in the later part of the 19th century. The first resulted in Alfred Hewitt's and Lorimar Fison's *Kamilaroi and Kurnai* (1883), a landmark study revealing the complexity of Aboriginal society, while pioneering new methods and ideas in anthropology including territorial organisation and rights determined by patrilineal descent. The collaboration between Hewitt, a bushman and explorer, and Fison, a

minister of religion, reflected growing public interest in documenting Aboriginal culture during the era of evolutionary anthropology. The second partnership was between Walter Baldwin Spencer, an academic biologist, and F.S. Gillen, the Alice Springs postmaster, which began during the Horn Scientific Expedition in 1894. Their collaboration involved three extensive field trips. Books published following each expedition would have a significant impact on ideas of social evolution. Important contributions were made to understanding the origins of Aboriginal art and ceremonial activities.

The German missionary Carl Strehlow, who spent thirty years in charge of the Lutheran mission in Hermannsburg near Alice Springs, opposed Spencer's evolutionary ideas. Both Spencer and Strehlow studied the Aranda people, but reached vastly different conclusions on their origins, which stimulated international argument. Strehlow challenged Spencer's evolutionary ideas on the basis of studies of language. This approach played an important part in establishing social anthropology. The collaboration between Spencer and Gillen continued until Gillen's death in 1912. Carl Strehlow's monumental studies of the culture and language of Aranda people were continued by his son Theodore Strehlow over four decades, to create the most comprehensive collection of Aboriginal culture.

Reverend Lancelot Threlkeld was the first to document an Aboriginal language comprehensively. He published the results of his study of the Awabakal language of the Lake Macquarie region in 1834.

Ministers of religion were well placed to record information on Aborigines and, for the most part, were well motivated.

There were also contributions from squatters and professionals who were in contact with Aborigines. Robert Brough Smyth (1830–89) was attracted by gold and eventually became a mining administrator. He took particular interest in Aboriginal protection as secretary to the Board for the Protection of Aborigines in 1860, and documented Aboriginal culture in *The Aborigines in Victoria* (1878). Edward Curr's long experience in rural Victoria and interest in the culture and welfare of Aborigines was summarised in 1886 in his four volume work, *The Australian Race: Its Origins, Languages, Customs.* Through his collection of artefacts he documented the lives of Aborigines in colonial Victoria.

Similar contributions came from explorers, notably George Grey who during his exploration of coastal Western Australia described many artefacts and important rock art. This long phase of documentation culminated in the quasi-professional studies by Herbert Basedow. His twenty years of careful documentation of ceremony, spiritual life, mythology and tribal organisation over a wide geographical area brought not only new information of local importance, but allowed the beginning of more generalised concepts, and new ideas on subjects such as totemism. A summary of his research was published in 1925 in *The Australian Aboriginal*, a watershed year in Australian anthropology. Basedow saw anthropology as part of natural history, and was an evolutionist influenced by his early studies in Europe on bones and skulls, though much of his fieldwork encompassed social issues. He recognised the importance of protection of Aborigines and their way of life. His doctorate included support for Thomas Hurley's "Austral Caucasian" concept of a shared common ancestry for Aborigines and Europeans. Basedow was a true polymath in the Australian context.

In 1914 the Australasian Association for the Advancement of Science invited the British Association for the Advancement of Science to hold its annual meeting in Australia. The anthropology section brought together German "static biological models" with Anglo-American ideas of a "more dynamic biogeographical notion of culture". This latter idea of adaptation and balance and the influence of environment brought anthropology broadly within a biological framework.

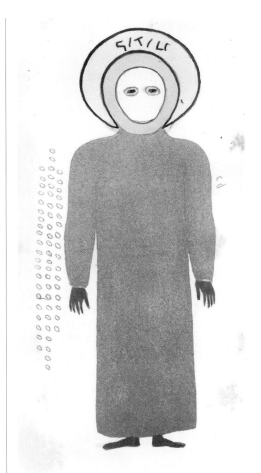

"Figure Drawn on Roof of Cave" in *Journals of Two Expeditions of Discovery*, G. Grey (1841)

Many of those who explored Australia made observations on Aboriginal people. Grey was amongst the first and most interested, experiencing a wide range of encounters from being speared to being saved from death — he even learnt an Aboriginal language. His two expeditions into the Kimberley region achieved little but distress, except for his recording of 4,000 year old Wandjina paintings representing cloud and rain spirits, and his discovery of probable descendants of Dutch shipwreck survivors.

Spencer pithily labelled it a shift from "kultur to culture". In this context, it is of interest that the Polish-Austrian anthropologist Bronislaw Malinowski, stranded in Australia by World War I after attending the same meeting, took the opportunity for "long-term deep immersion" amongst isolated native groups in the Trobriand Islands, to develop a model that would influence social anthropology.

When Papua was mandated to Australian administration in 1906 and when the Federal Government consolidated administrative control of the Northern Territory to South Australia in 1911, an impressive group of anthropologists took advantage of new opportunities to establish quality fieldwork studies in the two territories, often in association with administrators or missionaries. By and large they were valued by government as part of management plans for native populations. Most combined interests in natural history, ethnology and language with a focus on social anthropology. For example Norman Tindale (1900–93) who worked out of the South Australian Museum and developed the important idea of local boundaries of specific Aboriginal territories at a time when the belief that Aborigines were nomadic prevailed, while Arthur Capell (1902–86) combined anthropology with his interest in linguistics in Australia, Papua and Oceania.

In 1925, the year Basedow's book *The Australian Aboriginal* was published, A. Radcliffe-Brown arrived to take Australia's first chair of anthropology at the University of Sydney. Radcliffe-Brown championed the scientific method and was a strong supporter of the German-American Franz Boas, who promoted the idea that variations in behaviour are due not to characteristics selected in biological evolution, but rather that cultural differences are due to experience through learning, and that culture develops historically through people's interactions and the diffusion of ideas. Professor Radcliffe-Brown is considered the father of modern social anthropology in Australia, with much of his own work and ideas coming from fieldwork involving study of Aboriginal societies. He founded a framework of structural functionalism which he intended as an attempt to explain social stability, an inherently complex phenomenon given the variables that might be expected to overcome it. Radcliffe-Brown would influence a cohort of young anthropologists including Donald Thompson (1901–70), Ian Hogbin (1904–89), Ralph Piddington (1906–74) and W.E. Stanner (1905–81).

A.P. Elkin followed Radcliffe-Brown at Sydney University, where he became the major figure in anthropology in Australia for many years. He combined his academic role with a vocation as an Anglican minister. This vocational bent brought a sense of social justice and humanism to his signature attention to detail, and control

and compromise in his efforts to protect Aboriginal populations. Elkin's research focussed on ritual and kinship, while his theoretical construct combined elements of social anthropology and evolution. Many of his students and staff would continue his observational anthropology including Phyllis Kaberny (1910–77), Camilla Wedgwood (1901–55), Olive Pink (1884–75) and Ursula McConnel (1888–1957).

After World War II came another sea change in anthropology. Social evolution and then functionalist anthropology were replaced by a less restricted, more humanised and individualised approach to Aboriginal societies emphasising historical change. Leading this shift was William Stanner, who more than any anthropologist would shape the way Aborigines came to be seen late in the 20th century. Stanner's early influence was Radcliffe-Brown, but his particular contributions were summarised in the 1968 Boyer Lectures, "After the Dreaming" in which he introduced the idea of "the great Australian silence" and his term "a cult of disremembering". Stanner was credited for moving anthropology from a "melancholic footnote" in the book of Australian history to a central place where it underpins the humanity of Aboriginal Australians and the need for research.

Today a very different view is taken with respect to Aboriginal knowledge and to the use of a scientific method to understand indigenous peoples. Just as this text develops the idea that Enlightenment-driven Western scientific method was "owned" by all men and women in Colonial and post-Federation Australia, we now acknowledge a similar circumstance amongst Aborigines, enduring for thousands of years. Here the underpinning idea is land environment – and survival. The principals of empiricism and experiment that were applied in Australia before European settlement are essentially the same as those which guided white Australians. We recognise food management such as eel farms, fishing strategies and animal control, location through astronomy and cartography, and art, and a range of effective natural medicines. Aboriginal observers worked out how to predict changes in environment and weather by studying animal (including ant) behaviour. A comprehensive review of their traditional science would complement the current presentation, which is restricted to Western contributions.

A different twist on Australian influence in anthropology can be seen in the contributions by Raymond Dart (1893–1988) to understanding human evolution by the discovery of the first *Australopithecus africanus* fossil – an extinct hominan closely related to humans – in Taungs in South Africa in 1924. Dart was born in Queensland and studied medicine in Sydney. He was guided and influenced by James Wilson in Sydney and Grafton Elliot Smith (in London) to develop a career in anatomy, which became influenced by his fossil discovery. His idea of small brain, human-like posture and teeth, and location in Africa was at odds with concepts of human evolution at that time which envisaged large brains followed by humanoid characteristics – and in Europe! Eventually Dart's ideas were accepted with

"Colossal Brow-ridge, Arunndta man", in *The Australian Aboriginal*

palaeontologist Robert Broom stating Dart had made "one of the greatest discoveries in world history". Dart continued studies to show the tool making capacity of these "ape-men", generating further heated debate.

Important Anthropologists

Walter Baldwin Spencer (1860–1929)

Walter Baldwin Spencer was an English academic with a broad interest in zoology within a Darwinian framework, shaped by his time at Oxford University with the evolutionary biologist and ethnologist Professor H.N. Mosely. He was appointed to the foundation chair of biology at Melbourne University in 1887, where he would contribute widely to community and university life. A man of great energy, he participated in the Horn Scientific Expedition in 1894 as zoologist. This expedition and a meeting in Alice Springs with a talented amateur anthropologist, Francis Gillen, changed Spencer's professional and personal life by introducing him to social anthropology. He felt a deep affection and concern for Aborigines, recognising that "no sooner does the savage come into contact [with white men] than the changes in life . . . and introduction of diseases serve rapidly to bring about deterioration with young men bred from the wholesome restraint of the old men."

Anthropology fitted neatly into a zoological frame for an evolution-driven zoologist, with Spencer maintaining his global interest in zoology throughout his career in Melbourne. He took over responsibility for the Melbourne Natural History Museum in 1899, moving it from the University to the city, where he built a major anthropological collection. He maintained a personal and professional relationship with Gillen until Gillen's death in 1912. They made several major field trips leading to a series of books describing the characteristics and customs of Aboriginal people. Spencer was Australia's academic instigator of scientific anthropology, albeit within an evolutionary mould.

Herbert Basedow (1881–1933)

It must have been difficult for Herbert Basedow to know at any one time whether he was medical practitioner, geologist, administrator or explorer; but his abiding interest from his expedition with L.A. Wells in 1903 as a member of the South Australian Government North-West Prospecting Expedition, was anthropological study of Aborigines.

His summary of his findings published twenty-two years later, *The Australian Aboriginal*, states that he "lived among the uncontaminated tribes to study Australian anthropology at the fountainhead". While anthropology was gathering momentum driven by Darwin's natural selection ideas, Basedow's various callings brought him opportunity to observe "habits, laws, beliefs and

"Magic Implements" in *Across Australia*, B. Spencer and F. Gillen (1912)

Spencer was recruited to the Horn Expedition as zoologist, but finished as an anthropologist. He met F.J. Gillen at Alice Springs while on this expedition, a relationship that would develop into one of the most influential in Australian anthropology. The illustrations chosen reflect Spencer's initial interaction with Aborigines on the Horn expedition, and later studies with Gillen which were recorded in books, each covering a particular expedition.

legends" which he rightly perceived to be at risk of being "doomed to rapid extinction".

His appointments as Chief Practitioner of Aborigines and Special Aborigines Commissioner for both state and federal governments reinforced those opportunities, while his medical skills led to his acceptance by Aboriginal medical men as a kata or colleague. These relationships enabled him to witness rituals and sacred ceremonies, which were explained to him. Thus, he brought facts to light from new areas, in matters such as tribal information, initiation ceremonies, religious ideas (little of which was known at the time) and meanings in art. This included new insights into phallic worship and totemism, and connections to secret Tjuringa symbols.

Through medical experience he identified, more clearly than before, the extent of tuberculosis, syphilis and trachoma in Aboriginal populations, and together

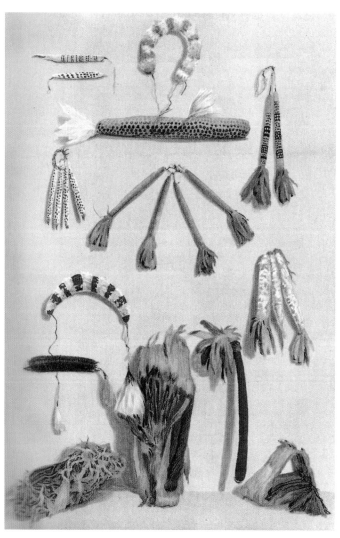

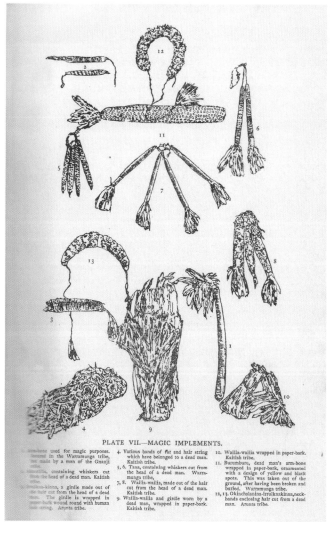

PLATE VII.—MAGIC IMPLEMENTS.

used for magic purposes, in the Warramunga tribe, made by a man of the Gnanji tribe.

containing whiskers cut from the head of a dead man. Kaitish tribe.

...-kinna, a girdle made out of the hair cut from the head of a dead man. The girdle is wrapped in paper-bark wound round with human hair-string. Arunta tribe.

4. Various bands of fur and hair string which have belonged to a dead man. Kaitish tribe.

5, 6. Tana, containing whiskers cut from the head of a dead man. Warramunga tribe.

7, 8. Wailia-wailia, made out of the hair cut from the head of a dead man. Kaitish tribe.

9. Wailia-wailia and girdle worn by a dead man, wrapped in paper-bark. Kaitish tribe.

10. Wailia-wailia wrapped in paper-bark. Kaitish tribe.

11. Burumburu, dead man's arm-bone wrapped in paper-bark, ornamented with a design of yellow and black spots. This was taken out of the ground, after having been broken and buried. Warramunga tribe.

12, 13. Okinchalanina-irrulknakinna, neck-bands enclosing hair cut from a dead man. Arunta tribe.

with other evidence of European occupation, led to his pleas for political intervention to preserve the health, welfare and culture of Aborigines in central Australia. As an advocate for Aboriginal people his chief purpose was to change their conditions. Basedow on his own initiative spent time in Europe before World War I honing his skills in the very new discipline of anthropology. He briefly studied in Breslau with Herman Klaatsch, the leading physical anthropologist of the time, who had spent time in Australia in 1904 (and who infamously sent a mummified body of an Aborigine back to Berlin). Klaatsch was the high priest of evolutionary anthropology at that time, with Basedow a close follower. Klaatsch in a major study of comparative anatomy purported to link Aborigines and neanderthaloids.

Basedow used his time in Europe to visit and study collections there of Aboriginal skulls and skeletons. The results of his studies – including studies of soft tissues in the field – were geared to contemporary evolutionary theory. For example conclusions such as the brain being "a little less complicated – than we are accustomed to see in our own sort" have no credibility today.

Basedow was the leading authority on Aborigines after the death of Baldwin Spencer, and with his passing anthropology moved from its unilinear evolutionary phase to the new phase of social anthropology, based largely on study of Australian Aborigines.

Alfred Radcliffe-Brown (1881–1955)
Bronislaw Malinowski (1884–1942)

The shift by anthropologists away from brutal evolutionary ideas which guided Spencer and Basedow became evident in the 1920s. Professional anthropologists

The Buffalo Hunter's Triumph", in *The Australian Aboriginal*, H. Basedow (1925)

Trained in medicine, geology, and evolutionary anthropology, Basedow progressively focussed his attention on the health, wellbeing and culture of Aborigines in central and northern Australia. The book from which these illustrations are taken is a classic summary of his observations, which focus on anthropometrics and were embedded in the then overarching science of evolution.

appeared and anthropology units were formed in institutes and universities. The two anthropologists credited with the rise of social anthropology, Alfred Radcliffe-Brown and Bronislaw Malinoswski, developed their ideas after studying social parameters in Australian Aboriginal populations and (in conjunction with the Australian government) islanders in the New Guinea archipelago.

Radcliffe-Brown studied Aboriginal populations in northwest Australia for two years beginning in 1910, and later took the foundation chair of anthropology in Sydney from 1926 to 1931, publishing his influential work *Social Organisation of Australian Tribes* in 1931. He began an era where observation and belief were replaced by strict scientific method. His focus on the whole society rather than the individual, studied somewhat dispassionately, brought him into conflict with colleagues, including Daisy Bates who had accompanied him in 1910. His claim that Bates was a disorganised amateur committed to the welfare of Aborigines, rather than to gaining knowledge, led to her empirical findings being discounted until recent times.

Radcliffe-Brown believed that anthropology could only be recognised as a branch of natural history if strict scientific methods were used. He was a big picture anthropologist and founder of Structural Functionalism, although he did not use that term. He believed that any society is a complex system whose parts worked together to promote the whole. For Radcliffe-Brown, all features, customs, practices etc. act in concert to contribute to a stable cohesive system. He postulated that stateless "primitive" societies lacking centralised institutions, are based on an association of "corporate-descent" groups and that these fundamental units of anthropology were processes of life and interactions. These processes are in a state of flux and the question he needed to answer was, in such circumstances, how was stability maintained? He introduced the idea of "co-adaptation" by which separate practices support each other to attain stability. This enabled scientific analysis of components to identify similar patterns in different societies in order to formulate and validate social symptoms by the systematic use of comparative methodology. The objective was to identify a "law of social development" to provide anthropology with a generic underlying principal.

Radcliffe-Brown's critics pointed to his neglect of the impact of historical changes, such as colonialism, on the development of a society, yet his ideas were to initiate the next important period of anthropological research. Radcliffe-Brown rejected unilinear evolutionary pathways, and the idea that a process of diffusionism underpins behaviour interaction between societies, because both point to studies of historical reconstruction rather than objective comparison, and were not susceptible to scientific interrogation.

Radcliffe-Brown shared his place as founding father of social anthropology with the Polish born ethnologist, Bronislaw Malinowski (1884–1942), who remarkably was studying nature groups in our region at the same time. Stranded in Australia at the outbreak of World War I (like Radliffe-Brown), Malinowski negotiated with the government for permission to continue his studies in the Trobriand Islands. By contrast to Radcliffe-Brown, he developed the idea of "participating observation" by merging with his subjects. His book *Argonauts of the West Pacific* became a pivotal document in the history of anthropology. His attention to detail and the individual (again contrasting with Brown) led to his thesis that societies develop to serve basic needs, a ground breaking concept that focussed on individuals with little understanding of or interest in how their society was shaped and where there were no guiding principles that would otherwise be expected to have framed an ordered institution. His idea that one had to "grasp the native's point of view" predicted modern views of anthropology, placing the ideas of individuals on their society as the pivotal dynamic.

Adolphus A.P. Elkin (1891–1979)

A.P. Elkin was in the right place at the right time when Alfred Radcliffe-Brown unexpectedly resigned his chair at Sydney University. Elkin was appointed Professor of Anthropology at Sydney in 1933. Australian born and an ordained Church of England priest, he had passion and commitment that drove him to research and document Aboriginal ways of life "before it was too late". With a passion for social justice and concern

The Southern Cross, from Groote Eylandt, in *Records of the American and Australian Scientific Expedition to Arnhem Land* (Vol.1 Art, Myth, Symbolism) Ed. C. Mountford (1955).

Like European observers of the night sky, Aboriginal artists borrowed from the animal kingdom in formal representations of stars, including this bark painting.

for the plight of Aborigines, he lobbied at every level of government for equality and protection. His views on integration made him a controversial figure in Australian politics. Elkin "drew upon ideas of cultural progress and social anthropology to propound a form of assimilation wherein the attainment of citizenship could be reconciled with the retention of Aboriginal identity and cultural distinctiveness" (Russel McGregor in *Oceania* 69 (1999) 243).

The strength of his research is his attention to detail in recording Aboriginal ways of life. He did not follow his predecessors' ideas on scientific social anthropology, rather he was more a social evolutionist and a believer in the ideas of diffusion with respect to acquisition of behavioural characteristics. For many years he was Australia's leading anthropologist (holding the only chair between 1934 and 1951), organising field studies and influencing a raft of assistants and students, while remaining a powerful political mover and shaker.

Frederic Wood Jones (1879–1954)
Charles Mountford (1890–1976)

Two very different men, Jones and Mountford came to anthropology from their "day jobs". Jones from academic medicine pursued in Britain, and Mountford an electrician working in the Postmaster General's Department. Mountford had left school at ten to work on the family farm in South Australia. Both were

interested in natural history and anthropology, each having been influenced by key encounters in their careers. For Jones it was working on skeletal remains in Nubia with the Australian Professor Grafton Smith which initiated his interest in evolutionary anthropology. For Mountford it was meeting Jones after Jones took the Thomas Elder Chair of Anatomy in Adelaide in 1920.

Jones guided and encouraged Mountford to record and analyse Aboriginal rock art, which he would continue until his last expedition in 1964. There was a quite remarkable similarity to the interaction between Francis Gillen, the postal station master at Alice Springs and Baldwin Spencer, zoologist cum anthropologist from the University of Melbourne. Jones became intensely interested in Aboriginal culture and wellbeing following his move to Adelaide, with his evolving anthropological interests based on his belief in Lamarckian evolutionary theory, always within the context of his academic focus on structure-function relationships. He established a successful anatomy department in Melbourne in 1930, retiring to London in 1938. After World War II he rebuilt the largely destroyed Hunterian Museum at the Royal College of Surgeons. His career is aptly summarised in *The Australian Dictionary of Biography*: "no one discovery or theory can be advanced [but his] contributions are impressive in the aggregate."

Seen as an amateur by some academics including

Elkin, Mountford brought great energy to numerous expeditions into interior Australia, documenting and collating Aboriginal art and culture as a social anthropologist. He left a collection of 13,000 photographs to the South Australian State Library, but he is better known for communicating Aboriginal ways of life through film. He became a trusted government advisor. In 1964 he submitted a masters thesis, "Ayers Rock, its People, their Beliefs, and their Art". His profile was international, which underpinned a successful application for funding from the US National Geographical Society to support the last of the great multidisciplinary scientific expeditions, the 1948 American-Australian Scientific Expedition to Arnhem Land. It was an ambitious and significant expedition involving a team of seventeen men and women led by Mountford over a nine month period. It focussed on documentation of the indigenous population and ecological issues involving fauna and flora at a new level in Australia. A fitting culmination for an extraordinary career!

Dorothy Sombono's "Bush Medicine Painting" in *Traditional Bush Medicines. An Aboriginal Pharmacopeia*, A. Barr, J. Chapman, N. Smith and M. Beveredge (1988)

The illustration of a tree called "Gardenia" is included in an extensive well researched pharmacopeia of plants found by observation and experiment over thousands of years to have medical benefit. The leaves are heated and crushed, and applied to relieve inflammation.

Chemistry

Chemistry is the branch of the natural sciences that includes the study of the properties, structure, composition and reactions of substances.

Initially this was about naturally occurring substances, but as scientists developed methods to synthesise substances and develop processes for commercial production of useful substances, the concept of chemistry broadened. The discipline of chemistry encompassed all these aspects in the first 150 years of European Australia, from optimising chemical production processes of minerals and agricultural products, to the analysis of Australian natural products and establishing chemical production facilities to counter geographical isolation, through to establishing a competitive presence at the cutting edge of international chemistry. Along the way John Cornforth was awarded a Nobel Prize. So was Robert Robertson, whose time in Sydney as Professor of Organic Chemistry was important to his work on natural products. While chemistry in colonial Australia was about practical survival, it was an exciting time of discovery in modern chemistry in Europe.

While convicts cleared land in Sydney Town, Antoine Lavoisier was busy in Paris establishing ground rules for modern chemistry. Before he was executed by guillotine in 1794, for collecting taxes in his day job, Lavoisier introduced precise quantification into chemistry, and by accurately measuring reactants, used it to prove conservation of mass, and to show an identity in principal between respiration in living organisms and combustion. He began the scientific process of classifying elements, breaking tradition to focus on "non-reducible elements". A century of progress beginning with Lavoisier let to the Periodic Table, created when Dmitri Mendeleev organised sixty six known elements into order based on

their atomic mass and in columns based on chemical similarities. Our period of interest is bookended from a chemical point of view by Crick and Watson in 1951 describing the DNA double helix with base pairing as the chemical basis of inheritance, to begin the modern era of molecular biology.

Through this period, Australia was directly affected by chemical discovery overseas, especially in Germany. Aniline from coal tar was the basic material for dyes and by the end of the 19th century German coal tar companies were developing nearly all the new coal tar dyes and synthesising or extracting ammonia, chlorine, sulphuric acid and many carbon compounds used for a range of civilian and military purposes. In 1828 when Frederick Wöhler synthesised organic urea from inorganic substances, he put paid to Vitalism, prevalent at that time, and began the science of organic chemistry. According to Vitalism, life depends on a force distinct from chemical and physical phenomena. If so it should not be possible to synthesise organic substances such as urea from non-organic material. Justus von Liebig identified in the 1840s the essential role of plant nutrients, with immediate relevance to Australian agriculture. In 1860 Gustav Kirchoff and Robert Bunsen developed spectroscopy as a tool for chemical analysis, a discovery of immense importance in identifying elements not just on Earth but on light emitting bodies throughout the Universe.

The capacity to identify elements in the Galaxy would supply critical information on the structure and events relevant to the origins of the universe. Important

Left: Alkaloid structure; right: Cholesterol structure

These structures of molecules central to the chemistry of plants and animals were identified by two Australian Nobel Prize winners, Robert Robinson and John Cornforth. Each had significant impact on therapeutics.

contributions began to come from Canberra's Mt Stromlo Observatory around 1920. Study of electromagnetic waves and the discovery by Wilhelm Röntgen in 1895 of x-rays generated from "discharge tubes", as well as the linked but accidental discovery of radioactivity the following year by the Frenchman Henri Becquerel, would have immediate impacts on Australia. The Professor of Physics in Adelaide, William Henry Bragg, began experimenting with electromagnetic waves, leading to his discoveries in x-ray diffraction, with the identification of the structure of crystals with his son William Lawrence Bragg. Father and son shared a Nobel Prize in 1915. Douglas Mawson published studies on detection of radioactivity in Australian minerals at the start of the 20th century, beginning a long chapter of contribution by Australians to atomic physics.

These were exciting times. The founding fathers of colonial Australian universities recognised the importance of the "new" science of chemistry (as well as the "old" science of physics) in the appointment of foundation chairs. In 1852 John Smith was appointed as Professor of Chemistry and Physics, occupying one of three Foundation Chairs at Sydney University. John Kirkland was appointed as the first Professor of Chemistry at Melbourne University in 1882. These were out of step with the explosive changes in contemporary chemistry in Europe, but the second round of senior appointments at both Sydney and Melbourne brought a different level of enquiry and organisation to shape the development of modern world class departments. These appointments were Archibald Liversidge in Sydney from 1872 to 1909, and David Masson from 1886 to 1923 in Melbourne. Great strides in establishing organic chemistry, with its applications to industry and the growing local interest in applying chemical analyses to natural products, led to the Sydney appointment of Robert Robinson (1912–15) to the first chair in organic chemistry.

These recruits from Britain were turning points in Australian science and all three would have profound effects. Liversidge and Masson established a scientific culture in their institutions and were important public scientists, spreading their culture through professional and social groups to promote the importance of science, teaching and research. They promoted the application of their discipline to the solution of practical problems, and in this fitted seamlessly into the Enlightenment influenced core of the Australian society that they had adopted. Much is owed to these men of science whose inheritance of the "Australian way" did so much to establish the connecting networks of chemistry and science in general in their adopted homeland.

Robert Robinson spent less time in Australia but left his mark in terms of establishing organic chemistry, and analysis and synthesis of natural products. He became a mentor and colleague to John Cornforth, who graduated from Sydney University and after a period of postgraduate research, left for Oxford on an 1851 Exhibition Scholarship in 1941 to work with Robinson on the structure of penicillin.

The period from the early 1920s to just after World War II consolidated the traditions in academic chemistry established by Liversidge and Masson. Their universities retained a focus on applied science, but more basic science was taught, with excellent up to date curricula supervised by young and enthusiastic Australian academic chemists with British training or experience. John Earl (1890–1978), was Professor of Chemistry at Sydney from 1928 to 1947, while Ernst Hartung occupied the Melbourne Chair from 1928 to 1952. Neither would make significant personal additions to scientific knowledge, but both developed good teaching programmes, especially Earl in Sydney. It was a difficult time because the Depression and World War II left little funding for research and teaching loads were considerable. It was also a time when significant postgraduate training in science was not available in Australian universities – there was no honours year added to a BSc degree and PhD training did not become available until the end of the 1940s. (The first Australian PhDs were awarded in arts in 1948, though science quickly became the main source of doctoral students. Today more than 100,000 PhDs have been awarded in Australian universities.) Those wanting a research career, would apply for a travelling scholarship, the most popular being the 1851 Exhibition Scholarships awarded to the best and brightest around the world, to do postgraduate study in British universities. Most of them would return to take leadership roles in Australia. Those

Leading academic chemists of the 20th century:

Scientist (and birth and death dates)	Under-graduate university	BSc	Location of post-graduate study	Academic roles and research fields
David Rivett (1885–1961)	Melbourne	1906	Oxford	• Professor of Chemistry, Melbourne • Head of CSIR (1926–49)
Frank Dwyer (1910–62)	Sydney	1931	Chicago	• Professor of Biological Inorganic Chemistry, ANU • Pioneer in metal coordination complexes
Adrian Albert (1907–89)	Sydney	1933	University of London	• Foundation Professor of Medical Chemistry, ANU
Rita Harradence (1915–2012)	Sydney	1936	Oxford	• Colleague and wife of John Cornforth Chemistry of penicillin and steroids
Arthur Birch (1915–95)	Sydney	1936	Oxford and Cambridge	• Professor of Organic Chemistry, University of Sydney • Biosynthesis of natural products including sex hormones
John Cornforth (1917–2013)	Sydney	1937	Oxford	• Professor of Chemistry, Medical Research Council, UK • Synthesis of non-aromatic steroids and stereochemistry of enzyme reactions
Ron Nyholm (1917–71)	Sydney	1938	University of London	• New South Wales University of Technology, then professor, University College, London Coordination chemistry; he initiated a renaissance in inorganic chemistry.
David Craig (1919–2015)	Sydney	1940	University of London	• Foundation Professor of Physical Chemistry, Sydney, 1952 • Theoretical chemist studying the nature of chemical bonds
Ron Drayton Brown (1927–2008)	Melbourne	1946	Kings College, London	• Foundation Professor of Chemistry, Monash University • Theoretical structural chemistry, microwave spectroscopy and galactochemistry
Jean Youatt (1925–2017)	Melbourne	1949	University of Leeds	• University of Melbourne • Pharmacological chemistry and chemical ecology
David Buckingham (1930–)	Sydney	1950	Cambridge	• Professor of Chemistry, Cambridge • Electric and magnetic properties of molecules and fundamental properties of matter

who did not (including world leaders in chemistry such as John Cornforth and David Buckingham) maintained contact with chemistry in Australian universities. The table includes many of those who were to make a difference in chemistry research, who were born and educated in Australia in academic departments run by second generation professors, themselves mainly Australian born.

During World War II many Australian chemists shifted their research to munitions and other areas of practical importance for the war effort. Since the 19th century, concurrently with the progress of academic chemistry, industrial chemistry had been playing an integral part in developing agriculture, grazing and mining – primary industries discussed elsewhere in this book, upon which national growth depended. Even applying known processes, such as treatment of phosphate rock with sulphuric acid to produce superphosphate and flotation to extract minerals, required the inventiveness and initiative that were bedrock characteristics of the colonial population. Academic chemists, like their colleagues in geology and biology, understood the importance of applying their skills to develop product from the unique materials found in their new land. Many papers appeared in late Colonial and post-Federation issues of the *Proceedings of the Royal Society of New South Wales*, where extraction and characterisation of substances, particularly oils, from native plants and trees were recorded by chemists, including Robert Robinson whose studies on natural substances would later win him a Nobel Prize. A testament to local innovation applied to the use of local products was production of gas from eucalyptus oil for lighting in the Victorian town of Kyneton in 1858. Important Australian commercial ventures applied industrial chemistry to maximise cost effective primary production, and to synthesis essential chemical products so isolation would not compromise Australia's progress.

Because of its technical base, the development of the chemistry industry in Australia was always linked to international activities. Many companies adopted the truncated structure of a branch plant operation dependent on overseas license. Others were to become subsidiaries of American or British corporations. Despite this, the chemistry industry in Australia pursued research more than others. Although industrial chemistry comprised only about ten per cent of Australian manufacturing, it accounted for about twenty five per cent of private sector research in the mid-20th century. The high cost of local research, and the small market, called for government protection. When protection was withdrawn in the second half of the 20th century, major multinationals including ICI, Monsanto and Union Carbide acquired positions in the Australian market. For example Timbrol merged with Eveready in 1957 to form Union Carbide (Australia). These international companies brought new technologies such as a switch

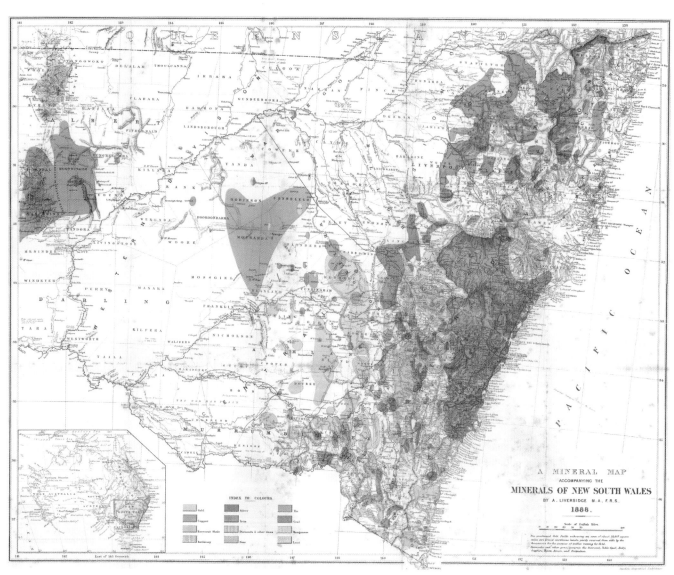

"A Mineral Map . . . of New South Wales", A. Liversidge in

The Minerals of New South Wales, A. Liversidge (1888)

Archibald Liversidge was a power at every level in late colonial science, while establishing a dynamic department of chemistry in the University of Sydney. His particular interest was the chemistry of natural minerals. His *Minerals of New South Wales* presents descriptions and analysis that he performed himself, to establish the prosperity of a colony so well endowed with mineral wealth.

in raw materials from derivatives of the gas and coke industries, to petrochemicals, and more cost-effective production, for example of ammonia for freezing plants and fertiliser.

Academics across a wide spectrum of disciplines until the mid-20th century focussed on applied science, so important to the progress and development of a young Australian society. Chemistry benefitted farming and the mining industries. Early entrepreneurs imported technology to produce soaps and glycerine, for example from fat obtained from abattoirs, and ammonia solvents from coal gas after coal firing plants began in 1841. Again and again an individual who saw a need and opportunity developed a strategy based on science to answer questions aimed at removing blocks to cost effective production. By making Australia self dependent they helped avoid the risk of being beyond supply lines at times of economic or military crisis.

The Colonial Sugar Refining Company (CSR), formed in 1855 to refine sugar from plantations along the North Coast of New South Wales, is an example.

When production and viability were struggling in its early years, Edward Knox was given the responsibility of making CSR efficient and profitable. Knox was not a chemist, but he believed in the power of industrial chemistry to optimise and monitor production processes and in the value of innovation. He recruited his first chemist in 1879 and by the time of World War I CSR employed about a hundred chemists. CSR was able to compete on equal terms with overseas producers, which surprised observers who were not aware of the progressive power of industry.

Another example is Timbrol, which produced chemicals from coal tar for medicine and industry – again the vision of an innovative individual. John Griffith Peake recognised the importance of Australia being self reliant on chemicals essential for farming, forestry and mining. He established Timbrol in 1925. Similar levels of initiative and innovation involving cyanide extraction of gold and flotation concentration of crushed ore, contributed to the international competitive nature of the mining industry.

Important Chemists

Archibald Liversidge (1846–1927)

Archibald Liversidge was the backbone of chemistry in Colonial Australia. He established a framework for the study of chemistry within his institution and at a national level, that attracted two remarkable individuals who would be awarded Nobel Prizes for Chemistry. A colleague summarised his contribution with the statement that Liversidge "was certainly the greatest organiser of science that Australia has seen and surely no one in the country ever worked more unselfishly and with greater singleness of purpose than he to serve science for its own sake." (R. MacLeod, 2009)

Liversidge's abiding professional interest was in mineralogy and the economic chemistry of mining, a very comfortable fit with mainline science in Australia in his time. This interest facilitated a close working relationship with the then Professor of Geology, Edgeworth David, as the two shaped a research-orientated university culture. Liversidge's many scientific communications included analysis of minerals and other natural substances, from gold to minerals in seawater. His main contribution to scientific knowledge was *The Minerals of New South Wales*, published in three editions between 1882 and 1888.

His most durable achievement was the organisation of science and the establishment of scientific networks within the University. Against much opposition from those in the classics, he created a Faculty of Science, and through his presence and position on the University senate, he helped drive a culture of enquiry essential for progress of the discipline. His chemistry teaching was of the highest standard, establishing an academic legacy that influenced many future leaders in the University and in economic chemistry. His first class in 1872 had ten students and was taught in two rooms. On his retirement, thirty five years later in 1907, there were 200 students with seven lecturers and demonstrators. His lecture notes (in the library of the Royal Society of New South Wales) reveal that contemporary chemistry at an exciting time of change was included. Expectations of students are reflected in the exam papers he set. In the year of his retirement, the Chemistry 3 exam included three papers: organic chemistry, inorganic chemistry, and history of the discipline. Recent discoveries including radioactivity, not yet a decade old, were examined in detail.

Liversidge promoted teaching of science in secondary schools. He took a leading role in the Royal Society of New South Wales, developing it to a new level. He regularly gave presentations on the chemistry of minerals, and his studies of nickel ore in New Caledonia encouraged development of a major export to Australia for processing. However, it was at an inter-colonial (then national) level that his efforts brought most reward. Liversidge recognised the importance of information exchange and debate over research – something sorely missing in colonial Australia. He aimed to make possible a serious debate in a communicating scientific community, from as early as 1880, when he unsuccessfully attempted to induce the British Association for the Advancement of Science to hold a meeting in Australia. Not dissuaded, and with enormous effort, he formed the Australasian Association for the Advancement of Science (AAAS), which held its first meeting in Sydney in 1888 – one hundred years after the First Fleet's arrival. A landmark of distinction for science, now with a unifying organisation across colonies, thirteen years before Federation! There were 850 who attended, which is remarkable given the cost and difficulty of travel in those times. He would remain secretary to the AAAS until 1909, after his retirement from the University. His vision following Federation was a new Australian academy of science, based in the national capital – a target reached in 1954.

David Masson (1858–1937)

David Masson's path resembled that of his Sydney counterpart, Archibald Liversidge. Educated in Scotland and with experience in Germany with Wöhler, the leading chemist in organic synthesis, he was recruited to the Melbourne chair of chemistry in 1886. Working with his academic colleagues Baldwin Spencer and Thomas Lyle, this group forged a solid academic base at the University of Melbourne for teaching and research in science. Teaching and administration loads limited Masson's personal research, though he continued studies in ion movement in solutions. He kept his relationships with international scientists especially William Ramsay in Bristol and was involved in discussions around the placement of argon, helium and radon in the rare gas section of the periodic table.

Masson's legacy to Australia was his contribution to the special societies and establishment of a broad framework for science in late Colonial times. He was a major contributor to the formation of AAAS, of which he became President, and the promotion of the Commonwealth Advisory Council in 1916, which led to the formation of the Council for Scientific and Industrial Research (CSIR). His successor to the chair of chemistry in Melbourne, David Rivett, was appointed Chief Executive of CSIR. Masson was a worthy successor to Archibald Liversidge as an organiser and promoter of Australian science.

Robert Robinson (1886–1975)

Robert Robinson was always a high flyer. He was awarded an 1851 Exhibition Scholarship to continue studies at Manchester University. Robinson was a towering intellect, taking the foundation chair of applied organic chemistry in Sydney in 1912 for three years. At this time chemists in Australia were particularly involved in extraction and characterisation of oils and chemicals from natural materials, often working with botanists such as Joseph Maiden. Robinson took advantage of access to botanical extracts from unique Australian flora, with a series of studies on eudesmin and its derivatives and on certain phenols extracted from eucalyptus species, publishing in the *Proceedings of the Royal Society of New South Wales*. Eudesmin was known as a constituent of folk medicines, and today is known to modulate cellular immunity. These were the first scientific studies in the chemistry of Australian plants and the foundation for a long series of studies including synthesis of organic compounds that would place Robinson at the front of synthetic organic chemists, leading to his Nobel Prize in 1947 for study of natural products including medicinal substances, alkaloids and dye stuffs. Robinson kept up his Australian connections, training many postgraduates, most famously John Cornforth, a Sydney graduate who would also be awarded a Nobel Prize.

John Cornforth (1917–2013)

John Cornforth was the only Australian born recipient of the Nobel Prize for Chemistry. Over and above his excellence in determining the steps leading to synthesis of the 19-carbon ring structure of cholesterol, by using radio markers to analyse placement of acetic acid subunits, and analysis of the stereochemistry of enzyme-substrate binding using asymmetry induced by isotope substitution, he contributed to the chemistry of penicillin production and in 1951 was the first to complete total synthesis of non-aromatic steroids. The impact and long term outcomes of his work continue today. Statins and new drugs blocking cholesterol synthesis are foundation stones in the prevention of atheroma-related disease in human beings. The influence of Robert Robinson, who built up the credibility of the chemistry department at Sydney University, attracted Cornforth to Robinson's department at Oxford, and the two established a collegiate relationship lasting many years.

Cornforth's opportunity came through experience in the Sydney University Chemistry Department, then under Professor John Earl. It was a modern quality department built on the contributions of Liversidge and Robinson. The contacts of the department were critical to Cornforth getting an 1851 Exhibition Scholarship to do a PhD with Robinson. Although John Cornforth never returned to work in Australia, he maintained close working relationships, and contributed through his support of ongoing development of chemistry in this country.

Edward Knox (1819–1901)
John Peake
Charles Potter (1859–1908)

Knox, Peake and Potter were men who understood the value of science to industry. By 1900 a major crisis in mining was compromising the Australian economy. Exploitation of the massive silver, lead and zinc deposits at Broken Hill had hit a road block. The superficial oxidised products were easily smelted. Not so the ores from the deeper lode, sulphides which being less amenable to economic extraction found their way to tailing dumps. Charles Potter, a Melbourne chemist and brewer, invented a flotation process by which crushed hydrophobic minerals, mixed up with waste matter, bound to air bubbles and floated to the top in an industrial bath, where they could be concentrated and separated out. D.G. Delprat and others on site improved the process by adding oils, making it possible to extract large volumes of ore that would otherwise have been wasted. Later improvements in efficiency included pumping air under pressure and adding xanthates to increase the hydrophobicity of mineral surfaces and make the extraction process more selective. Flotation, with its refinements, invented and developed in Melbourne and Broken Hill in the early 20th century, has been described as "the most important scientific invention in Australia of economic value".

The development of the flotation cell by Graeme Jameson at the University of Newcastle has revolutionised mining, especially of coal, showing the continued development of flotation principles into the modern era.

John Griffith Peake was an Australian engineer and chemist who had seen the need for an independent supply of chemicals essential to industry and medicine, following his experience in World War I and in England recovering from war wounds. In Sydney in 1925, with the financial help of Sir William Dixson, he began the industrial chemical company Timbrol to extract timber preservatives from coal tar obtained by the Australian Gas Light Company as a by-product of gas production. Timbrol began with relatively low value products obtained by distillation, fractionation and extraction technology. Products included cresylic acid (a germicide), naphthalene for mothballs, creosote, phenols and solvents. From the 1930s derivatives were synthesised, including aromatic hydrocarbons such as nitrobenzines and aniline for explosives and xanthates, useful for mineral flotation thanks to improvements in the extraction processes pioneered at Broken Hill. Later Timbrol developed a chloro-alkali plant to synthesise chlorohydrocarbons used as herbicides, fungicides and insecticides for agriculture.

In the later part of the 20th century Australian producers would be acquired by multinational chemical companies, such as ICI and Monsanto, after government protection ended and new overseas technologies left them uncompetitive.

Working with the Colonial Sugar Refining Company (CSR), Edward Knox (and later his son E.W. Knox) valued the use of scientific research at every stage of production. CSR's first chemist, Andrew Fairgrieve, was recruited from Scotland in 1879. By 1900 CSR employed fifty chemists, and fifteen years later the number had doubled. Trained chemists from Australia and overseas were employed but Knox preferred to recruit bright young school leavers for training on the job. Research into carbonation to clarify juice from the mills, and the use of "diffusion" to maximise extraction from sugar cane, were two of the research programmes that gave CSR an edge over local competitors and allowed it to compete internationally. By 1900 CSR produced 70% to 80% of Australian sugar, largely due to the work of chemists and to its excellent quality control processes. That CSR with cane could match the standards of German competitors refining beet sugar was "something that is not supposed to happen in a colonial setting".

David Rivett (1885–1961)

David Rivett was the essence and spirit of the Council for Scientific and Industrial Research (CSIR) – under his leadership from 1927 to 1946 it would become a highly respected organisation through which science based resolution of challenges to primary industry could be channelled. His youth in Tasmania in a poor but principled family, shaped the character Rivett built into the CSIR in the years before World War II – of honesty, integrity, courtesy and pursuit of excellence. He never lost sight of the bond between basic and applied science but brought a naivety untarnished by politics or personal gain. He had graduated in chemistry at Melbourne University under Professor David Masson, a respected mentor and friend.

After a stint at Oxford supported by a Rhodes scholarship, in which his work was linked to the wartime challenge of manufacturing ammonium nitrate for explosives, Rivett was convinced of the value of basic science in solving the problems of industry. It was a principle upon which he would later build the CSIR.

He replaced his role model David Masson in the Melbourne chair of chemistry in 1924 but was soon appointed to the CSIR executive committee, and then became chief executive officer. Making that move was a difficult decision.

When he found limited resources and funds were available, he focussed on a limited number of projects, funding studies in academic organisations across the country, and recruiting the very best young scientists. The 1930s brought advances in animal disease prevention, remedying trace element deficiency in animals and plants, meat chilling for export and timber processing.

David Rivett was an outstanding administrator who encouraged a "society of co-workers", taking a personal interest in his workforce. However the years after World War II were a difficult time for him – his views on openness conflicted with a prevailing habit of secrecy that came with the Cold War, he was losing his fight to avoid becoming a tool of secondary industry, and his views on independence and basic research made him a political target. He championed astrophysics, introduced new discussions in animal genetics, animal physiology, atomic physics and meteorological physics, but political intervention compromised the integrity and independence of the CSIR by moving towards a social agenda.

His position was replaced by a five member executive and the Public Service added new layers of bureaucracy to the administration, which imposed a public service mentality on the organisation. Rivett retired in 1949 to play roles in the development of the Australian National University and the Australasian Association for the Advancement of Science as well as taking several board positions.

In the CSIR (and CSIRO, which it became in 1946) Rivett was able to implement his vision of what was needed – acquisition of knowledge applied to the solution of industrial challenges, through building a world class institution focussed on applied science. He underestimated the power of politics but he anticipated the limitations that would bedevil the work of universities by supporting truly national organisations dedicated to research. It is no accident that the glory days of the CSIR and CSIRO were imprinted by his leadership.

Biomedical Science

Over the last century, biomedical science has been a strong suit of Australian science.

About two thirds of all Nobel Prizes won by Australians have been in the category of Physiology and Medicine, mostly for discoveries related to infection or immunology relating to the host-parasite relationship. This is not surprising considering that colonial Australia lived with ever present danger of infection and in fear of the "great European epidemics", smallpox and bubonic plague. Yet the enlightenment principals of initiative and innovation that drove other areas of scientific endeavour, were less obvious in biomedical science, for several reasons.

First, residents of colonial Australia were not worse off from a health viewpoint than their cousins in England. There was a similar life expectancy and background mortality. Many immigrants from Europe chose to come to Australia for its relatively healthy environment, including some who made major contributions to science such as William Farrer, John Ashburton Thompson, Ferdinand von Mueller and Joseph Maiden.

Second, doctors in Colonial times were not considered "men of science" by an educated colonial public. Admission into hospital was feared as much as epidemics of smallpox and plague. The choice between regular or unorthodox practitioners was based more on price than any scientific confidence in trained doctors.

Medical science had stood still for two thousand years, in a belief structure codified by the Greco-Roman physician Galen. This framework based on health equating to a balance between four humors, was an extension of the idea of Empedocles that all matter is made up of four elements.

As late as 1900, many doctors practising in New South Wales retained belief in the primacy of Galenic medicine, rather than the specificity that came with Pasteur's germ theory. In the forty years that followed colonisation, doctors were employed by the state, and work in medicine was part time, with much of their time given to administrative and commercial interests. From the 1850s professional groupings appeared, with

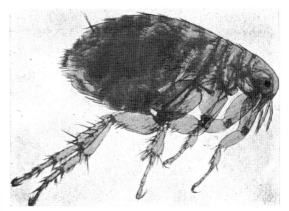

Plague Flea, a photomicrograph from Ashburton Thompson's 1902 report

The diseases that drove emotional reaction in colonial Australia, and thus led to change, were not those that killed most (like tuberculosis or influenza) but rather the "classical" or "great" plagues of smallpox and the bubonic plague. They had limited exposure in Australia but were pivotal in creating government and medical response in public health measures and research. The key person was John Ashburton Thompson, an epidemiologist whose focus and commitment essentially established public health on scientific grounds. His work on plague confirmed that the pathogen was carried by fleas to rats, and lead to the logical practices of elimination of rats and limited use of quarantine.

local medical journals that included new ideas and contemporary European discoveries.

Publication of medical journals began in 1844 when William Bland formed the Medical Chirurgical Association of Australia. More enduring journals began in Melbourne with the *Australian Medical Journal* in 1856, and in Sydney with the *Australian Medical Gazette* in 1881. By the end of the 19th century there were about 450–500 medical practitioners in each of Victoria and New South Wales.

It should also be mentioned that overseas scientific publications arrived with remarkable speed in colonial Australia, with journals such as *Nature* coming from Britain by the first available ship. In papers written on the bubonic plague epidemic in Sydney in 1900

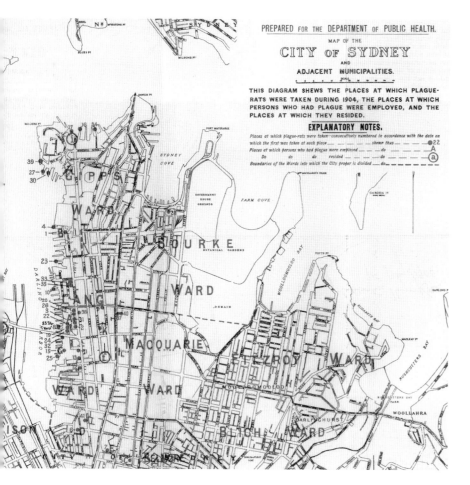

Plague Report (1904)

This map accompanied the report on the fourth outbreak of plague in Sydney. It includes the data identifying capture of culture-positive rats and the relationship of these to the reduction of human disease— a triumph in pandemic control by Ashburton Thompson.

Colonial doctors, new medical challenges were recognised and published. Thomas Robertson working in the Parramatta practice and described as the colony's first full time doctor, injected fluid from a lesion in a patient with black leg (now known as anthrax) into a kitten in 1851. Post mortem findings of splenomegaly and intestinal gangrene were typical of anthrax infection. The patient had been skinning sheep and anthrax (diagnosed then as Cumberland disease) would become a major threat to the pastoral industry. The recognition of Cumberland disease as anthrax and its prevention by immunisation introduced 30 years later by Louis Pasteur's nephew, would become a defining point in biomedical science. Robertson published his findings in the local newspaper, a common way of communicating scientific discovery at that time.

A second "new" disease in humans found by doctors working in rural Australia was hydatid disease. Hyatid disease in human beings is caused by ingestion of eggs of the tapeworm *Echinococcus granulosus* excreted by dogs that have eaten infected sheep offal. The parasite was introduced into Australia with sheep. The sheep-dog cycle has been brought under control, but a sylvatic (feral) form involving macropods such as kangaroos, and dingoes (first described by Bancroft in 1890), has become a significant problem. In the mid-20th century a survey in Tasmania showed one in a thousand people was infected with cysts, and that 500 to 600 new cases were diagnosed each decade, while 60% of sheep carried cysts and 12% of rural dogs were infected with adult worms. Commensurate with growth of the wool

by John Ashburton Thompson, related discoveries in contemporary reports from the Pasteur Institute in Paris are quoted. The international telegraph link completed in 1872 from Adelaide to the British network in Cornwall ensured immediate contact with events in Europe.

A departure in Australia from the rigid British structure of medical research meant the absence of a hierarchical system, with opportunities more reflective of the contemporary egalitarian society that gave short shrift to traditions. This would change somewhat after 1880 when the British Medical Association with its imperial agenda (like other British organisations such as the Church of England) established colonial branches in Australia, adding a political emphasis to the growing profession. One example of the benefit of "colonial freedoms" was the speed and enthusiasm with which discoveries became part of routine practice. Discoveries such as anaesthesia with chloroform or ether were used by Australian medical practices within months of their description in European or American journals.

While few significant discoveries were made by

industry, serious, often life threatening, hyatid disease was becoming a common problem in rural areas by the mid-19th century, and research into hyatid disease, its diagnosis and treatment, became a significant area.

The first major study in Australia was by John Davies Thomas, a Welsh surgeon working in Adelaide. He published two accounts detailing clinical features and surgical treatment of cysts. The second, edited and published in 1894, after his death, remained a definitive and exhaustive statement until the monograph published in 1928 by Harold Dew, professor of surgery in Sydney.

Joseph Bancroft who first recognised the sylvatic cycle of hydatids also identified a second parasitic disease. Bancroft was a remarkable man whose contributions from his base in medical practice in Brisbane between 1864 and 1894, added to knowledge in natural history, medical science and agriculture. His main contribution to medical science was his description of the nematode *Wuchereria bancrofti* as the cause of filariasis. His suggestion that transmission involved microfilaria and mosquitos was later validated by his son.

Biomedical science came of age in an Australian context after lean years of little achievement, in an explosive way, due to a coincidence of events in the late 1880s. Perhaps short of a perfect storm, but important nevertheless. Like other shifts in Colonial and post-Federation science, the change can be attributed to remarkable individuals: in this case John Ashburton Thompson, Adrien Loir and Harry Allen.

John Ashburton Thompson (1846–1915) was an English-born physician trained in the new science of epidemiology and public health who dragged a reluctant colonial government into recognising the importance of science in the control of infectious disease. His contributions are legend across an extraordinary range of public health problems, from tackling lead poisoning in children in Broken Hill to tracing an epidemic of typhoid fever to its source in a Sydney dairy and winning international recognition for his work on leprosy. But it was his scientific analysis and management of bubonic plague in Sydney in the early 1900s that would most influence the structuring of public health in Australia. Thompson adopted the Germ Theory of Louis Pasteur

with vigour and enthusiasm as a tool to study and manage infectious disease. He recruited Frank Tidswell as bacteriologist, whose studies did much to clarify the transmission of plague bacillus from rat to man via fleas. Tidswell's collaboration with Charles Martin exploited an 1895 development by Albert Calmette who had modified the ideas of Pasteur regarding immunisation in order to develop snake antivenoms.

Thompson's successor as advocate for scientific public health was Melbourne-born physician John Cumpston. Cumpston, who joined the Federal Quarantine Service in 1910, said he was "afire with enthusiasm for the new bacteriology, the new pathology and the new epidemiology [as] beacons indicating the new road to prevention of disease on a national scale," committed to continuing the foundations of public health begun by Thompson. He understood the contemporary view in Britain, that health research was a public responsibility, and that such support was basic to the establishment of centres of excellence in biomedical research. He became a major driver in 1927 of a Federal Health Council, which became the National Health and Medical Research Council a decade later.

Adrien Loir, by his very public involvement in practical aspects of the new biomedical science initiated in Paris by his uncle, Louis Pasteur, kindled the attention of Australians to the sea change occurring in biomedical science. Here was Pasteur's nephew touting a cure for the rabbit plague a decade on from Pasteur's critical studies establishing Germ Theory as causation of infection by growing the anthrax bacillus in culture to deny for ever spontaneous generation as the origin of infectious disease. Loir smuggled chicken cholera bacteria into Australia as a candidate pathogen for dealing with the rabbit plague, but what actually got attention was his recognition that black leg and Cumberland disease were one and the same: anthrax, for which Pasteur had an effective attenuated vaccine. He established a Pasteur Institute in Sydney in 1890, and produced the vaccine there that changed the course of a disease which threatened the Australian wool industry by causing over 400,000 sheep deaths annually. Loir established a credibility for biomedical science and initiated a focus of interest in host-parasite relationships that would be a dominant part of veterinary and medical research thereafter.

Harry Allen, a graduate in medicine from Melbourne University saw the limitations for research within academic departments due to many distractions that included teaching, administration and service loads. Committed to the idea of research, he promoted dedicated research institutes, independent but attached to academic and hospital centres. This was insightful, as it is very much how academic medicine has evolved. University medical departments would play a crucial role in identifying the importance of research, without themselves being cutting edge centres of research activity. Characteristics of medical schools that promoted research careers included a breadth of biology as part of a commitment to graduates having skills to cope with all challenges in remote areas, and a structured course along lines introduced by Abraham Flexner, with a two phase curriculum consisting of medical science blocks followed by clinical teaching in hospital-based departments.

Harry Allen was appointed to Melbourne University's first chair of pathology in 1882, twenty years after Professor George Halford's appointment, to begin the University's medical faculty. Allen was acutely aware of opposition by doctors at the Melbourne Hospital and the lack of resources and time for research. He knew of turf wars in Adelaide and saw the delays in establishing a medical faculty in Sydney, caused in part by those in the classics who opposed the University's entry into professional training. The one academic leader who shared Allen's passion to promote research was Professor Thomas Anderson Stuart, the foundation dean of medicine in Sydney when it began in 1883.

Stuart was recruited from Edinburgh with funds provided by the bequest of John Challis in 1880. He would remain in charge of the medical school until his death in 1920. Stuart was an imaginative and dynamic dean, recruiting quality academics he knew from Edinburgh, determined to establish a research culture within the faculty. His key recruit was James Wilson in 1887 as demonstrator in anatomy. Wilson's passion was comparative anatomy and the scientific method. It has been accurately stated that "Anderson Stuart built

the Sydney Medical School; Wilson furnished it with a reputation."

Stuart continued his inspired recruitment, appointing a young Charles Martin in 1891 as demonstrator in physiology. Martin and Wilson formed the core of an outstanding research group – Australia's first in bio-medical sciences – that they termed "the Fraternity of Duckmaloi". Grafton Elliot Smith, a medical student, joined the group which created a productive programme recognised internationally for work on comparative anatomy and physiology of marsupials and mono-tremes – an extraordinary achievement at that time.

Harry Allen and Anderson Stuart – representing rival medical schools – understood the need for research and the obstacles it faced. They worked together to support the creation of the first medical research institute, the Institute of Tropical Medicine in Townsville in 1910. The initiative and most of the funding came from local residents, concerned that medical problems unique to a tropical environment were not a priority for the medical schools in the south (though behind the scenes, Stuart supported attempts to move the Institute to Sydney). Anton Breinl was recruited from the Institute of Tropical Medicine in Liverpool, England. Despite opposition and numerous hurdles, Breinl assessed local needs and established a formula for medical research institutes, which continues today. It involves establishing research institutes within major hospitals, combining the provision of laboratory medicine as a service, with research. Indeed, the Townsville "experiment" began the discipline of laboratory science in Australian medicine. The formula was not always successful, because the provision of pathology services often compromised the scientific agenda.

Harry Allen was a key person in establishing in 1915 Australia's most successful institute of medical research, the Walter and Eliza Hall Institute of Medical Research (WEHI). Allen opposed views that would have restricted research strategies, seeing the importance of independence in terms of research agenda and connections with hospital and university, encouraging what he called "interactive independence". After a staggered start Charles Kellaway was made Director, a position he held until 1944.

Kellaway was the face of the WEHI, shaping a course and recruiting the best of scientific minds, to earn international recognition. He managed to secure the first Commonwealth grant of £2,500 for research into polio, hydatid disease and snake antivenom. This grant was a model for future funding by the National Health and Medical Research Council. Kellaway understood the importance of applied research and encouraged close cooperation with the newly formed Commonwealth Serum Laboratories. It was that interaction that led to the production of snake antivenom in 1930.

In the 1950s, many of Australia's leading scientists in infection and immunology were at the WEHI. Best known was Frank Macfarlane Burnet who spent most of his fifty nine year career there, serving as Director from 1944 to 1965. His contributions to the understanding of infection, especially influenza, were immense. Yet halfway through his directorship, in 1957, he switched the focus of the institute from infection to immunology. By then he had developed the understanding of immune tolerance for which he won a Nobel Prize, based on a differentiation between "self and non-self" in embryonic life, and produced ground breaking ideas on clonal selection which switched mainstream immunology from humoral to cellular immunology. The switch to immunology was not as fundamental as many thought because it simply involved a focus on a different aspect of the host-parasite relationship, i.e. the specific host response rather than the pathogen.

Another favourite son of the WEHI was Neil Hamilton Fairley, a graduate in medicine from Melbourne. He had an uncanny knack of collecting like-minded enthusiasts around him, and early colleagues included Charles Martin and Charles Kellaway. Kellaway invited Fairley to work at the Institute, on the snake venom project. In World War II, he played a central role in the control of tropical infections in the Pacific Theatre. His role in field studies of malaria led to atebrin being developed, a drug that would cause a dramatic fall in the incidence of malaria, making a significant contribution to the war effort.

Towards the middle of the 20th century, Burnet recruited a band of impressive young scientists, including Frank Fenner as a virologist, Gordon Ada in biochemistry and Ian Wood to run a new clinical research unit in 1946. Wood had worked with Fairley in the malaria programme and began blood banking on a tight schedule for the war effort in 1940.

Wood was the most complete physician I ever worked with. He developed a programme in chronic hepatitis, identifying chronic active and chronic inactive hepatitis. These studies would result in his protégé Ian Mackay describing autoimmune hepatitis – closely correlating the clinical focus with Burnet's switch from virology to immunology. Mackay would become internationally recognised as the "father of autoimmune disease".

The success of the first early institutes encouraged other major teaching hospitals to develop research institutes, usually integrated to some extent with pathology departments. In 1926 the Baker Institute in Melbourne began a long interest in metabolism and cardiovascular disease. Joe Bornstein developed the first assay capable of measuring insulin and identified for the first time type 1 and type 2 diabetes based on detecting available plasma insulin. In Sydney the Kolling Institute of Medical Research was opened at Royal North Shore Hospital in 1920. Its international reputation was established by Rudi Limberg's crucial studies on porphyrin metabolism. At Sydney Hospital in 1933 the Kanematsu Institute was best known for work in neurobiology when John Eccles was Director and Bernard Katz worked there. Both went on to win Nobel Prizes.

Charles Martin, from *The Martin Spirit* by Patricia Morison, (2019), p.109

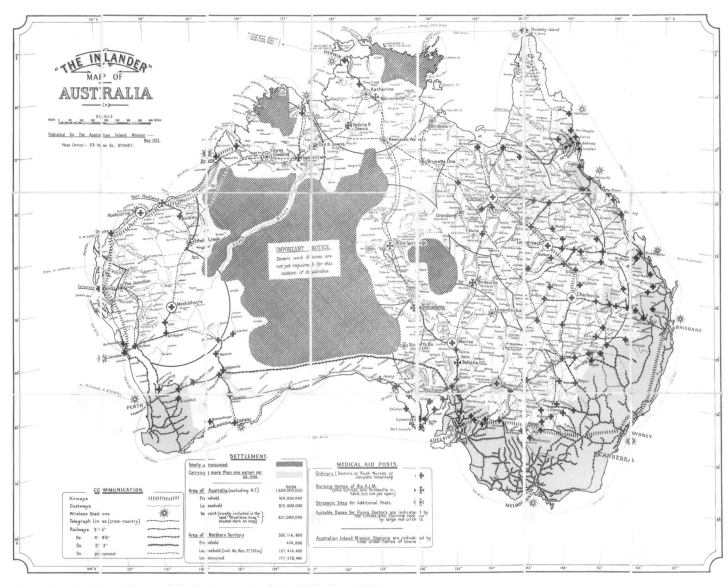

"The Inlander Map of Australia", the Australian Inland Mission, 1922

This map illustrates John Flynn's response to the immense challenge of providing medical services to outback Australia, including the Flying Doctor service to remote areas.

In Adelaide, the Institute of Medical and Veterinary Science was primarily a diagnostic service for the Royal Adelaide Hospital. In Queensland the Queensland Institute of Medical Research at the Royal Brisbane Hospital was begun in 1945, to focus on tropical and subtropical infections, involving Edward Dereck's studies on Q fever (including his discovery of *Coxiella burnetii*), scrub typhus, and leptospirosis, and Ralph Doherty's studies on dengue, typhus and leptospirosis in cane cutters.

By 1950, biomedical research was in a good place. Medical schools across Australia taught science-based medicine and for bright students, science careers were encouraged by funding available through a National Health and Medical Research Council grants programme. Academic departments with a research orientation were appearing across the medical spectrum and PhD programmes were available from 1948 to kick start academic careers. By 1966 the Council's grants had grown to exceed $1 million annually, from a base 30 years before of £30,000 (By the year 2000 grants awarded were in excess of $180 million). From three PhDs awarded in 1948, in 2011 in Australia there were 6,780, 25% of which were in health sciences.

Important Biomedical Scientists

Adrien Loir (1862–1941)

Adrien Loir introduced the idea of biomedical science into Australia during two stays in Sydney between 1888 and 1892. He brought the new science of bacteriology and immunology to Australia in a very practical and public way. Trained in Paris in bacteriology by his uncle Louis Pasteur, he was sent with vials of chicken cholera cultures, to compete for a £25,000 prize (more than $10 million in today's money) to be awarded for successful eradication of the billion rabbits literally denuding the Australian countryside. In the event, the pathogen was never field tested, because of a combination of bureaucratic intransigence, Pasteur's pride and inflexibility due to an inability to see past his laboratory results, and genuine concerns that native birds would be infected, and that the poultry industry could be at risk. Pasteur had been consolidating his Germ Theory over the previous twenty five years. He was winning the argument against a two thousand year entrenched dogma linking infectious disease to miasma and imbalance of the four humors, by demonstrating that each infectious disease was produced by its own specific pathogen.

Loir's contribution to colonial Australia was not to be bio-warfare against the profligate rabbit, but to demonstrate that an attenuated anthrax vaccine protected sheep and cattle against infection and to develop an economically successful immunisation programme. The Germ Theory was seen to have enormous practical implications for human and animal health. In 1891 he presented a paper to the Royal Society of New South Wales, "Notes on the Large Death Rate amongst Australian Sheep in Country Infected with Cumberland Disease or Splenic Fever." Cumberland disease was killing well over 200,000 sheep each year. Loir's detractors stated that 200,000 was not a significant loss in over 50 million sheep, but the problem was the distribution, not just the number. In some farms, annual loses were up to fifty per cent of their flock. The cause was uncertain and thought to be a toxic vegetable until Loir recognised the anthrax bacillus.

Loir completed a controlled trial of the vaccine in an affected property at Junee, New South Wales. He immunised twenty sheep which all survived, while sheep that were not immunised all died within forty eight hours of infection with anthrax. This made a dramatic impression on the farming community and Loir set up a production unit to make the vaccine on Rodd Island in Sydney Harbour. One hundred and thirty thousand sheep were involved in his programme to control a major challenge to the wool industry.

The impact of Loir's years in Australia was enormous – his attempts to introduce a pathogen to control rabbits and then his vaccine programme introduced the Germ Theory and contemporary European biomedical science. Everyone in the colonies knew of Adrien Loir and the Pasteur Institute.

Interestingly, Harry Allen was chairman of the rabbit commission. His immediate relationship with Loir was hostile and he opposed the field trials with rabbits using Loir's vaccine. Allen was, however, influenced by the new science and would become a major promoter of bacteriology in medical practice. Loir and John Ashburton Thompson were both contributing members of the Royal Society of New South Wales, closing a loop of communication between this triad of medical scientists.

Harry Allen (1854–1926)

It would surprise many to see Harry Allen identified as a significant figure in biomedical research, if they knew only of his autocratic rejection of Pasteur's plan to use chicken cholera as a pathogen for rabbit control or of his lack of research expertise.

Allen was a rigid and difficult man, not particularly popular amongst colleagues. His own biomedical research was not distinguished and led him to overstate the incidence of chronic syphilis in Melbourne. However he had remarkable administrative skills and understood better than others how to develop medical research. He had visited European cities and been exposed to the influence of Pasteur. He also recognised the difficulties of research from a traditional academic base. Despite the limits of his research contribution, he understood the importance of developing independent institutions where research is the primary purpose, yet saw advantages in linking these entities to hospitals and pathology departments, as well as their value to universities as a way for students to have access to cutting edge research. He was a major force in developing the first medical research institute, in Townsville, and Melbourne's Walter and Eliza Hall Institute which soon became a world class centre. He insisted on its independence when it came to research agenda. Amongst the clutter of academics who emerged in late Colonial times, Allen was one of few local medical graduates who provided leadership in academic medicine, which he maintained well into the 20th century.

John Ashburton Thompson (1846–1915)

A third foundation figure of scientific medicine in Australia is John Ashburton Thompson. His energy and focus dragged a struggling public medical service into becoming a highly organised machine based on scientific analysis and research. His efforts culminated in the establishment of the National Health and Medical Research Council and a scientifically based public health structure the equal of any in the world. He has aptly been called "a giant amongst pygmies" in reference to his eminence among his colleagues.

Thompson came from London for "his health", with unusual training for those times – in epidemiology and public health. He came in 1884, soon after smallpox arrived in Sydney. New South Wales had 227 cases between 1881 and 1885. Thompson's skill led to his appointment as medical officer and then as Chief Medical Advisor. Over thirty years he created a contemporary public health department served by highly competent scientists including Frank Tidswell, and William Armstrong (1859–1941) – who curbed infant mortality and, as Director General of Public Health, continued Ashburton Thompson's ideas of disease prevention through environmental control.

The scientific base for development of public medicine is Thompson's legacy. The two great epidemics that struck real fear into the colonial population were smallpox and bubonic plague. Thompson established his reputation and began his career in public health in Australia by applying his expertise to smallpox in the early 1880s, and his career culminated in his application of scientific principals in the control of bubonic plague at the turn of the century.

The world's third great pandemic of bubonic plague began in China around 1890 and continued for fifty years with at least 10 million deaths. Thompson knew the danger to Australia and followed the discoveries by Yersin, Simond, Haffkine and Hankin out of Pasteur Institutes in Bombay and Vietnam through his subscription to the *Annales de l'Institut Pasteur* in the two or three years before plague was diagnosed in Sydney. Review of the epidemics in Sydney in 1900, 1902 and 1903 illustrates the quality of Thompson's scientific approach and his success in neutralising an unparalleled medical threat to the people of Sydney.

The epidemic in 1900 had 303 patients. One hundred and three of them died of the disease. It was associated with terror, pandemonium, and major disruption made worse by an inflammatory press, a confused government and numerous offers from quacks. Services were closed, health teams took punitive measures and an illogical quarantine process was implemented. Thompson with his bacteriologist Frank Tidswell examined 15,000 rats and their fleas and confirmed Simond's controversial idea of 1898 that the rat flea was the vector of infection and that human infection followed epizooic infection in the rat, man then becoming an "accidental host". Rejecting Hankin's public health concept of "place infection" based on person-to-person transmission of disease, Thompson believed the sole focus should be on eradicating the source – the rat population. In the 1900 epidemic a hysterical public and government did not listen to Thompson, with chaotic outcomes and a community cost estimated at £176,000.

Thompson was not just a scientific physician, but a consummate communicator who worked to educate decision makers in the value of science-based decision making. His brother, a journalist, was a valuable political ally. In 1902 when a second plague epidemic visited Sydney, Thompson controlled the response by eradication of the rat using only limited isolation of the patient. Disruption was controlled, quarantine limited, and cost contained at £24,000, and only 139, less than half the previous number, were infected – a remarkable demonstration of the value of scientific medicine.

Thompson was convinced he could predict any further epidemic, by monitoring infection in the rat population of the Woolloomooloo to Darling Harbour arc, aimed at detecting any epizootic in local rats. A surveillance programme in 1902–03 studied 31,000 rats. In 1903 1.25% of rats and 0.83% of mice were found to be parasite positive. A full blown epidemic was prevented by an intense de-ratting programme following recognition of an infected rat, and only two cases of bubonic plague were diagnosed – a remarkable and unprecedented outcome. Thompson's discoveries and subsequent strategies are likely to have contributed to Australia being the only continent without an infected native rodent population.

James Wilson (1861–1945)
Charles Martin (1866–1955)
Grafton Smith (1871–1937)

Wilson and Martin were British medical graduates, each with a passion for research, and both recruited to the faculty of medicine at Sydney by the foundation dean, Thomas Anderson Stuart. They formed the basis of a group that attracted outstanding students, especially Grafton Smith, and collaborators including James Hill (a Scottish biologist recruited to Sydney as a demonstrator, who became a leading embryologist and returned in 1906 to the chair of anatomy and zoology at University College London).

Wilson was the group leader – a lifelong naturalist who established a department of anatomy along Edinburgh lines, with an outstanding culture for research. Together with colleagues at Sydney University, Archibald Liversidge and Edgeworth David, Wilson established a scientific research culture in a university that had historically favoured classics. Through their involvement with the Royal Society of New South Wales, the Linnean Society, the Australian Museum and the Australasian Association for the Advancement of Science, he promoted academic science in Australia. In the University he was the catalyst for the first informal

group of scientists focussed on a research theme that was interdisciplinary and self supporting. It was a model taken up by few in those times. The main theme was study of the anatomy and physiology of native fauna. Wilson's study of the embryology of the platypus (with James Hill) was internationally acclaimed. More than anyone, Wilson was responsible for the development of a research-based academic department within the faculty of medicine and he was widely recognised for establishing the reputation of the Sydney medical school. He took the chair of anatomy at Cambridge in 1920. In Sydney he was followed by his protégé John Hunter (a prodigious local medical graduate who brought scientific acumen to neuroanatomy and neurophysiology, until he died of typhoid at twenty six. He was revered by colleagues and students).

Charles Martin's time in Australia was brief, from his arrival in Sydney till he left in 1903 after a stint at Melbourne University, to take the directorship of the Lister Institute in London. Like John Hunter, Martin gained profound respect and influence in medical research. In Sydney he established a close relationship

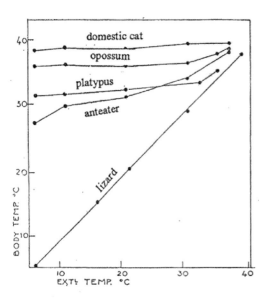

Diagram on body heat control, from *The Martin Spirit* by Patricia Morison, (2019), p.43

Martin was the very essence of medical research in Australia. After investigating marsupial physiology, including temperature regulation, and snake venom antidotes, he returned to England as Director of the Lister Institute which would train many of the leading medical research workers of the 20th century.

with James Wilson, studying the physiology of mono-tremes and marsupials including heat regulation in the platypus. He also worked with Frank Tidswell on snake venom actions and the bubonic plague.

Martin was an inspiring leader in medical research who never lost his affinity with and love of Australia. The Lister became a mecca for young Australians seeking research opportunities and experience maintaining a powerful influence on Australian research. Those young Australians included Thomas Dunhill (a Melbourne surgeon who revolutionised thyroid surgery), Howard Florey, Neil Fairley, Macfarlane Burnet, Gordon Cameron and Charles Kellaway. The most prestigious College of Physicians travelling research scholarship for young physicians, is named after Charles Martin. Though thought of as a physiologist and pathologist, his broad interests included host-parasite relations in infections, especially enteric fever. He developed the TAB vaccine which was produced by the Lister Institute, when he found enteric fever at Gallipoli was caused by paratyphoid A and B microbes. He returned to Australia to lead the Division of Animal Nutrition in 1930 at the CSIR for three years. Back in Cambridge his field studies of myxoma virus predated its use to control the rabbit plague in Australia.

Grafton Smith was influenced by Wilson during his undergraduate medical training, becoming a demonstrator in anatomy a year after graduation. There he became an active research member of the team headed by Wilson, and gained an MD for studies on the structure of the brains of non-placental mammals. His research continued overseas – initially at Cambridge for nine years, then in Cairo, followed by chairs of anatomy in Manchester then London. He revolutionised the teaching of anatomy, and always maintained contact with science in Australia. Many Australians trained under him including John Hunter and Joseph Shellshear, who both made important contributions to anatomy in the Sydney medical school. He developed an interest in anthropology following his time in Egypt developing the diffusionist school of thought with respect to cultural migration and was an important promoter of the first chair of anthropology at Sydney University. He had a profound impact on Raymond Dart whose later discovery of *Australopithecus africanus* would revolutionise thinking on human evolution.

This remarkable group of medical scientists, together for but a few years, would provide a model that influenced the way science, especially anatomy and physiology, developed through the 20th century.

Howard Florey (1898–1968)
Roy Cameron (1899–1966)
Henry Harris (1925–2014)

Three Australian graduates in medicine who moved to England for training in experimental pathology stayed on to become world leaders in biomedical research. Howard Florey was awarded a Nobel Prize for his work with Ernst Chain in developing penicillin as a therapeutic, jump starting the antibiotic era. Roy Cameron contributed basic understanding of regeneration of islet cells in the pancreas and liver after damage related to the pathogenesis of cirrhosis.

Henry Harris was a consummate cell biologist who showed that latent virus infection could induce cell fusion, which then become a tool to study control of gene expression in normal cell function and in cancer, including the first clear demonstration of tumour suppression genes.

These three giants of experimental pathology had immense influence and were recognised with awards and titles at the highest level. Throughout their working lives, they attracted postgraduate students from Australia and in different ways maintained contact with Australia, particularly by providing research training. Harris left Sydney to do a PhD with Florey at Oxford, while Cameron maintained interest and influence through colleges of pathology. Australians who trained in experimental pathology with Cameron included George Christie and Don Wilhelm, who respectively would chair academic departments in Melbourne and Sydney. Florey became involved in the idea of the Australian National University's John Curtin School of Medical Research between 1947 and 1957, essentially becoming its non-resident head. In 1965 he became Chancellor of the University.

What characterised each of these great experimental pathologists was the breadth of their work and their far reaching influence. Howard Florey's appointment to the Sir William Dunn Chair of Pathology in Oxford was a milestone. For the first time a man who saw pathology as disordered physiology was appointed to a chair of pathology, a position he held from 1935 to 1962. He was succeeded by Henry Harris, giving an Australian continuity at the head of one of the most prestigious departments in pathology over nearly sixty years. While Florey is best known for his work in converting Alexander Flemming's observation of 1928, that penicillium mould secretes an antibacterial substance, into the practical antibiotic penicillin, he also contributed in studies of inflammation. In particular he studied lysozyme in surface secretions as a natural antibiotic.

As Florey had a broad grasp of the interface of physiology with pathology, so Harris grasped cellular and molecular genetics, including the role of nuclear and cytoplasmic RNA in message expression and mapping of gene loci on chromosomes.

Roy Cameron was more in the mould of a classical experimental pathologist, but again his experience and interests covered the questions of the day in anatomical pathology. The sweeping vista of their professional interests was the basis of their extensive contributions to biology and pathobiology, of a nature rarely seen in today's world of specialisation.

Charles Kellaway (1889–1952)
Frank Macfarlane Burnet (1899–1985)
Neil Fairley (1891–1966)
Ian Wood (1903–86)

Charles Kellaway was a Melbourne medical graduate who fell into research following World War I, working at the interface of physiology and pharmacology, influenced by the famous pharmacologist Sir Henry Dale. Kellaway's early work expanded and clarified understanding of anaphylaxis. He then used this experience to analyse the pathophysiology of snake envenomation. His best known research was the discovery of slow reactive substance of anaphylaxis (SRS-A), an important mediator of allergic reactions. His personal discoveries were not as important to medical research as his role in making the Walter and Eliza Hall Institute the leading biomedical research institute in

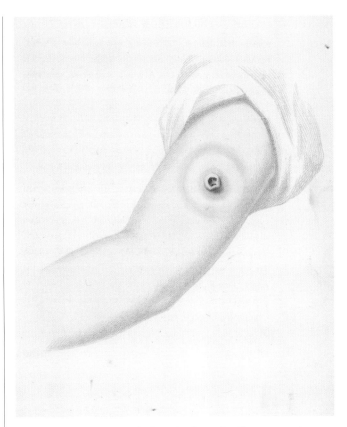

"Varicella vaccine" in *An Inquiry into the Causes and Effects of the Variolae Vaccinae,* E. Jenner, Government Printer, NSW (1800 — second edition, 1884)

the country. He achieved that by recruiting and then creating career paths for the best young scientists and by shaping the Institute around infection as an underlying principal relevant to health.

Kellaway maintained connections with the Melbourne Hospital, the University of Melbourne and the Commonwealth Serum Laboratories. He fought to ensure cashflow and government grants. His work with government established the idea of government responsibility for funding research, a concept well established in Britain at that time.

The heartbeat of the Institute, whose imprint made it an international leader, was Frank Macfarlane Burnet, described by his successor, Gus Nossal, as "a contemplative, almost solitary kind of genius". Burnet's career was a constant flow of ideas. Not in the mould of a textbook leader, Burnet by sheer brilliance inspired loyalty and cohesiveness simply because bright men and women wanted to be part of his team.

The first phase of his career related to infection,

with important contributions across the board illustrating his broad grasp of biology. He spent twenty years in the new field of virology, particularly with bacteriophages – viruses infecting bacteria – and then on influenza. His early work on bacteriophage biology showing integration of viral genes with the host genome was a prelude to the later development of microbial genetics. He became best known for his influenza virus work, beginning with the development of techniques to grow the virus in fertile hens' eggs, the basis of commercial mass production for the purpose of vaccine manufacture. He worked with colleagues including Alfred Gottschold and Gordon Ada to define the biology of infection and the broad genetic apparatus and novel methods of gene re-assortment that are the basis of genetic shift – the phenomenon that produces a constant challenge to those attempting to produce a contemporary vaccine.

Burnet's group showed RNA to be the primary genetic information and characterised the two surface proteins responsible for adhesion and penetration. His work with the polio virus showed more than one type, a finding of central importance to subsequent vaccine development. Broad interest in infection led to his identification of the rickettsial pathogen that caused Q Fever (a serious zoonosis of abattoir workers). He identified the cause of the tragic deaths of twelve children in Bundaberg in 1928, following their immunisation against diphtheria. When he injected rabbits with the inactivated toxin produced by staphylococci that he found to be contaminating the diphtheria vaccine (which had caused the deaths of the Bundaberg children) he noted a progressive increase in level of antibody response to repeated injection.

This was the observation that changed Burnet's (and the Institute's) agenda from parasitology to immunology, from the pathogen to the host's response. The anamnestic antibody response to repeated antigen injections suggest memory requiring a cell-based response, flying in the face of the prevailing view that the antibody was moulded by the antigen, an instruction theory of antibody synthesis. Such a theory would not predict an antibody response increasing over time, which rather was more consistent with cell division. An unexpected experiment of nature, in this case a disastrous one with contaminated vaccine, triggered an important insight that would change the way scientists think. It initiated a period of fifty years of research in cellular immunology, begun by Burnet at the Walter and Eliza Hall Institute.

Later demonstrations by Nils Jerne that natural antibody occurs in the absence of any antigen exposure confirmed Burnet's view that the instruction theory was incorrect. In 1960 he was awarded the Nobel Prize for his discovery of acquired immunological tolerance, which he first discussed in a monograph published in 1949, "The Production of Antibodies". Burnet's recognition that the differentiation of self from not-self was developed in embryonic life was a prelude to a new era, based on immune tolerance and rejection and the clonal selection theory. The Walter and Eliza Hall Institute's change of direction in the mid-1950s reflected the leap and insight that would lead the world in the new arena of cellular immunology. Immunology research had been all about antibody. Burnet's clonal selection theory, by focussing on antibody as a cell-surface receptor, dramatically shifted the focus from humoral (antibody) to cellular immunology, that would dominate immunology research until the development at the end of the century of whole animal concepts using "knock-out" mice.

Burnet understood the importance of applied research yet recognised the dangers of clinical medicine suffocating research initiative. He established a collaborative venture with Ian Wood – a fine physician who had developed blood transfusion services and worked with Neil Fairley in World War II on malaria and penicillin. Following the War, Wood became Assistant Director of the Institute and developed the Clinical Research Unit, where many future leaders of medicine in Australia would work. When Burnet switched from virology to immunology, Wood followed suit. His clinical unit had focussed on gastroenterology where he had invented the first flexible gastric biopsy tube (the first specimen was taken from Wood himself showing gastric atrophy and, unrecognised at the time, pernicious anaemia). Wood recruited Ian Mackay who arrived as Burnet was announcing the switch in focus to immunology. Autoimmune disease became the theme of the Clinical Research Unit following the discovery that many patients in the Unit presumed to have chronic viral hepatitis had positive autoantibodies and some had a positive test for lupus erythematosus cells, the classic autoimmune marker for systemic lupus erythematosus (SLE). Burnet's and Mackay's book on autoimmune disease became the bible establishing a new era and the beginning of clinical immunology as a medical discipline. Ian Mackay stands out as the father of autoimmunity.

An early appointment to the Institute was Neil Hamilton Fairley, a local medical graduate who would have a profound influence on both the future directions of the Institute, and Australian medical research. His career came about through contacts made during service in World War I, where he met Charles Martin and Charles Kellaway. Following the War, he worked with Martin who was Director of the Lister Institute in London. Martin and the Lister were a rite of passage for many young Australian medical graduates including Frank Macfarlane Burnet, looking at a research career. With training in tropical medicine, Fairley joined the Walter and Eliza Hall Institute under its first director Sydney Patterson in 1920. The focus on infectious disease suited Fairley, whose first projects were on syphilis and hydatid disease. As part of this, he documented the incidence of parasitic infection at the local abattoirs. His energy and commitment had a major influence on Burnet whose earliest memories at the Institute were of working with Fairley on antibody responses in typhoid fever. It was that work which led to Burnet's first forays into virology via bacteriophages infecting the typhoid bacillus and related bacteria: the beginnings of microbial genetics, and some would say of molecular genetics.

Fairley balanced his interests in research and tropical medicine with time in Bombay and London, before returning to the Institute. Influenced by Martin's work on snake venom, he worked with Kellaway to produce the first commercially available snake antisera in Australia. This public coup signalled a commitment by Kellaway, as Director, to applied research with the newly formed Commonwealth Serum Laboratories.

Fairley came into a new phase of his career during World War II, when his work on dysentery and malaria made a significant difference to outcomes in the Pacific theatre. Malaria was an enormous concern because the War was being fought in malarious areas, the Japanese had captured most supplies of quinine, and prevention was then based on the impossible task of avoiding mosquitos. The Land Headquarters Medical Research Unit was established and based in Cairns under Fairley, to discover an alternative prophylactic regimen. A candidate was atabrine developed in Germany. However, the level of protection and effective dosage were not known. Twenty thousand anopheles mosquito larvae from New Guinea were flown in each week to feed on infected soldiers, then used to infect more than 800 volunteers in the largest clinical studies involving transmission of infection to be done. The discovery of atebrin as an effective solution is an outstanding example of experimental clinical medicine.

Frank Fenner (1914–2010)

An Australian medical graduate who had once wanted to be a geologist, Frank Fenner established his credibility as a virologist with landmark books *Medical Virology* and *The Biology of Animal Virology* that consolidated a broad understanding of the fledgling science of virology. Less well known was his co-authorship of *The Production of Antibody*. That was the book which first identified the self versus non-self concept of antibody response, and the idea of non-responsiveness or tolerance, that led to Burnet's award of the Nobel Prize. Fenner joined Burnet at the Walter and Eliza Hall Institute after contributing to the control of malaria in World War II. Within two years at the Institute, Fenner had established a reputation in "pox disease" with his work on the pathogenesis of ectromelia or "mouse pox".

There followed a short period at the Rockefeller Institute in New York studying an atypical mycobacterium that had been isolated from patients with skin ulcers in Bairnsdale, that became known as Buruli ulcers. He studied the biology of the bacteria and noted a temperature effect on their growth which restricted lesions to the skin. From the Rockefeller he was recruited to the foundation chair of microbiology at the new postgraduate university in Canberra where he would continue for the rest of his career. He picked up the opportunity to continue working on pox viruses with an epidemic of released myxomatosis concerning health professionals. Over fifteen years, he would complete studies on every aspect of myxoma virus infection in rabbits including molecular and genetic studies on the emergence of resistance to infection as the mortality in nature declined, based on the first studies of recombination in a DNA virus. From his years of study of pox virus in animal models, it was a short jump to be given the chairmanship in 1967 of the World Health Organisation Global Commission for Certification of Smallpox Eradication. An important part of the programme was to prove that monkey pox differed from human smallpox, in an era before gene sequencing was routine. In a sense, this safety check continued an earlier challenge, proving that rabbit pox did not infect human beings. In that case, he had simply injected himself with the rabbit pox virus. Twelve years later Fenner was able to announce that smallpox was human specific and had been eradicated from the Earth. It was an achievement of immense importance in medical history, the culmination of a career focussed on a global understanding of pox virus infection. Eradication of smallpox was an outcome of application and lessons from scientific study of the disease process in animal and human models dating back to Fenner's war experience with malaria.

A humble and unselfish man, Fenner was dedicated to peace and our environment. No finer role model exists for an aspiring young scientist.

Anton Breinl (1880–1944)
Edward Derrick (1898–1976)
Joseph Bornstein (1918–94)
Max Lemberg (1896–1975)

The Walter and Eliza Hall Institute set the pace and standards for biomedical research institutes in Australia – it established a framework for success that no other completely reached. The importance of

From left to right: Macfarlane Burnet, John Eccles, Howard Florey

Three powerhouse Nobel Prize winners in medical science in the mid-20th century were Macfarlane Burnet, Eccles and Florey, recognised on stamps for, respectively, establishing cellular immunology, clarifying transmission across neurological synapses, and developing penicillin to initiate the age of antibiotics. They shaped the continuing strengths of Australian biomedical science in the study of host-parasite relationships and in neurosciences.

its "detached relationship" with a major hospital, its trueness to an underlying theme, focussed leadership and political support, combined to marshal success. The same effective principles to varying degrees fell short with other institutes established before 1950. The recruitment of John Eccles and Bernard Katz – future Nobel Prize winners – to the Kanematsu Institute at Sydney Hospital was an opportunity lost through failure to recognise and support the independence of research agendas: a blueprint on how not to develop a medical research institute.

The title of first Australian medical research institute goes to Townsville. Anton Breinl was the first director there of the Australian Institute of Tropical Medicine, formed to better understand how Europeans could live in tropical conditions. Politics underpinned its beginnings and its end in 1930, when it was argued that the Institute had satisfied its primary goal of showing that "a working white race" could perform strenuous labour in tropical conditions without bad outcomes. In reality Anderson Stuart, Dean of Medicine at Sydney, had always wanted the Institute to be part of the Sydney Medical School. It was not a coincidence that it was "moved" to Sydney in 1930.

Breinl was recruited from the Tropical Medicine School in Liverpool, where he had an outstanding research career that involved the demonstration in 1909 that arsanilic acid (Atoxyl) could cure trypanasomiasis – the beginning of chemotherapy and the inspiration for development of Salvarsan to treat syphilis by Nobel Prize winning Paul Ehrlich in Germany. In Townsville, often with little support, Breinl began surveys into patterns of disease in northern Australia and developed laboratory support that was incorporated into hospital medicine – the first clinical biochemistry diagnostic unit. Breinl described unusual tropical medicine and patterns of disease and pioneered studies of physiology and biochemistry of Europeans in a tropical climate. He published twenty two papers – mainly in *The Australian Medical Journal* – during his twelve years in the Institute. It is disappointing that politics and lack of support (plus a touch of racism during World War I provoked by his Austrian origins) limited the career options of a man described by Paul Ehrlich as "one of the leaders of modern chemotherapy".

Twenty years after Breinl's resignation, Edward Derrick, an Australian born doctor with experience in the Walter and Eliza Hall Institute and in London, would rekindle the idea of a tropical medicine institute where the diseases occurred. A man of strong religious convictions who returned to Australia as another "health migrant" to recuperate in a healthy environment from tuberculosis, he settled in Brisbane. He became interested in fevers and clinically defined Q fever in abattoir workers and isolated the pathogen. Collaborating with his old colleague Macfarlane Burnet, Derrick characterised the rickettsia organism, which was named *Coxiella burnetti*. He showed that it was transmitted by ticks and that bandicoots were a reservoir. He also studied a number of leptospirosis subtypes, and was the first to isolate *L. pomona*. In 1944 he successfully lobbied for the start of the Queensland Institute of Medical Research – essentially to resume the work of the Townsville institute.

The Queensland Institute opened in 1947 with Derrick as deputy director. He became director in 1961. The primary focus was study of febrile disease in tropical areas, particularly in leptospirosis, ricketsial disease, dengue and scrub typhus. Derrick developed an interest in viral disease, which would become a strong

suit of the new Institute. Later he pursued research in epidemiology of asthma and would publish over 120 papers on local infectious disease. The Queensland Medical Research Institute became a strong mid-ranking international research institute maintaining its focus on infection, and building an unrelated interest in cancer.

Joe Bornstein's discovery of the essential lesion difference between type 1 and type 2 diabetes in 1949 highlights the research contributions of the Baker Institute. It was begun by John Mackeddie in 1926 as an adjunct to Melbourne's Alfred Hospital, "to improve laboratory facilities and keep up with research". These worthy goals were a little short of what is needed to create a world leading research institute. The central focus became cardiovascular disease and related factors such as diabetes, hypertension and obesity. Research encompassed a wholistic approach encompassing exercise and drug treatment and extending from birth to death. Bornstein had come to Australia as a child, and trained in medicine, and joined Basil Corkhill's diabetes research team. He developed a complicated bio-assay using alloxon treated diabetic rats that were hypophysectomised and adrenalectomised to test insulin levels in plasma from type 1 and type 2 diabetes to demonstrate a dramatic difference between diabetes with and without available plasma insulin. This was published in *The British Medical Journal* in 1951, a decade ahead of the radio-immunoassay results that earnt Rosalyn Yalow a Nobel Prize. Bornstein would be appointed foundation Professor of Biochemistry at Monash University. Diabetes research continued within the Baker Institute, integrated into vascular research.

Wilson Ingram (1888–1982), a clinical pathologist trained in medicine at the University of Aberdeen, was the founding director of the Kolling Institute at Royal North Shore Hospital. His career bridged clinical and laboratory medicine with a particular contribution to the management of diabetes. His theatres of contribution included two world wars and the BANZARE 1929–31 expeditions to Antarctica with Mawson, which he served as doctor and biologist. The Kolling Institute was closely tied to the Hospital, and never evolved a consistent theme. It became known through the work of Max (Rudi) Lemberg – who escaped Nazi Germany in 1933. He was appointed by Ingram to the position of Director of Research Biochemical Laboratories. Rediscovery of cytochromes (which had been forgotten) and the planar structure of the porphyrins that they contained led Lemberg to study the binding of the iron complexes of porphyrins to protein in oxygen carriers, enabling an understanding of the importance of conformational changes in the associated protein in the control of the central iron of haematin enzymes such as cytochrome oxidase. He wrote the definitive text on haematin compounds, establishing himself as a major research worker in tetrapyrrole biochemistry. His work continued on structure–function relationships of porphyrins, the structure of porphyrin prosthetic groups and the biosynthesis of porphyrins. Rutherford Robinson stated: "Lemberg gave a distinctive and intellectual stature to the science of biochemistry in Australian equalled by only a few."

Australia's biomedical institutes have yielded many scientific contributions, yet perhaps only the Walter and Eliza Hall Institute has consistently and to theme provided a style and international leadership throughout its existence. Much credit belongs to those few who led and set a framework for its evolution.

John Eccles (1903–97)
Bernard Katz (1911–2003)
Stephen Kuffler (1913–80)

John Eccles was a Melbourne medical graduate who began his career in research through a Rhodes Scholarship in 1925, working with the British neurologist Charles Sherrington on reflexes and excitation and inhibition. This became a broad brush template for subsequent studies on transmission across synapses in both the central and peripheral nervous systems. In 1937 he took the directorship of the Kanematsu Institute at Sydney Hospital where he collaborated with Bernard Katz and a research student Stephen Kuffler, continuing work extending the hypotheses of ionic mechanisms of membrane transmission which included inhibiting synaptic action and the "sieve hypothesis".

Much of this work involved insertion of micro-electrodes into nerve cells – a method developed by Eccles and his team. From the molecular level, he moved to understand functional organisation within the brain, and the workings of the structural patterns formed by particular aggregations of nerve cells in different anatomical locations. In later life, Eccles like many great scientists examined the philosophy of biological events. As Burnet focussed on ideas of aging, Eccles focussed on what he saw as "whole brain science". His later work was at the Australian National University, before he moved to Chicago and Buffalo in the USA.

Bernard Katz joined Eccles at the Kanematsu Institute in 1938, having graduated in Medicine in Germany and completed a PhD in England in neurophysiology. He would discover watershed moments in neurophysiology using the intracellular recording method developed by Eccles to monitor endplate potentials at the neuromuscular junction, to demonstrate that neurotransmitting materials are released from presynaptic vacuoles in discrete quanta. He identified "miniature endplate potentials" due to a background spontaneous release of acetyl choline and their sensitivity to osmotic pressure and calcium concentration.

The work of Katz and his colleagues published in the *Journal of Physiology*, and their insight into the release of neurotransmitters were rated on par with Hodgkin's and Huxley's studies on ionic changes in neural transmission. In Sydney, he began work on synapses that would lead to the award of a Nobel Prize in 1970. The stimulus for much of this work was interest in nerve gasses containing organophosphates which block ganglionic transmission. Bernard Katz became an Australian citizen and fought in the Pacific War. His work with Eccles on the structural-functional relationship in nerve-muscle synapses was a foundation stone to the electrophysiological studies that followed. His subsequent career at University College in London was stellar.

The convergence of the three neurophysiologists at the Kanematsu was a critical convergence of research pathways in a new type of synergy of laboratory and academic cooperation, with its focus on the neuromuscular junction as a central element in understanding the function and structure of the nervous system. It was the first time an Australian institute attracted world class international scientists to train with an Australian scientist. With time that kind of international focus would become a core value of the Walter and Eliza Hall Institute.

The achievement of the team at the Kanematsu was due to a synergy that included Stephen Kuffler, who collaborated there with Eccles and Katz. Later recognised as "father of modern neuroscience", Kuffler was a medical graduate from Vienna, with a passionate focus on scientific method in searching for the truth. He did groundbreaking work on synaptic transmission, begun in Sydney, and retinal function. He discovered the primary inhibitory neurotransmitter in the brain. His vision and energy established in Boston the first neurobiology programme in world research.

Norman Gregg (1892–1968)
John Cade (1912–80)

Norman Gregg and John Cade were medical graduates whose discoveries in the 1940s would change thinking in two important areas of science. Both were highly intelligent clinicians who thought outside the rigid medical box of those times by observing and collecting data to move medicine in new directions. Because their discoveries were slow to be recognised internationally, Australian scientists had the best opportunities for further development.

Gregg's observation that infection with rubella in early pregnancy is teratogenic challenged existing belief that birth defects are inherited and the placenta is an impenetrable barrier to infection. C. Swan followed Gregg's discovery by polling doctors to find that two thirds of women infected with rubella during pregnancy had babies with birth defects. Twenty years later William McBride's recognition of the teratogenic potential of thalidomide dramatically challenged the prevailing view of drug safety in pregnancy and established what has become a major plank of clinical medicine – the teratogenic effect of drugs in pregnancy. Australians were the prime movers in environmental influence on teratogenic outcomes of pregnancy.

Cade's observations that lithium salts suppress mania and act as a mood stabiliser (probably even today, the only true mood stabiliser), turned clinical practice on its head. Psychiatry had been dominated by Freudian psychodynamic theory involving a conflict between fundamental structures of the mind. Cade's discovery introduced the idea of chemical lesions. The concept underpins modern psychiatric practice, but was not accepted internationally until the 1970s. The hiatus was filled by research by Australian scientists, especially studies by T. Nowak and W. Trautner on the physiology of lithium treatment including its specificity, kinetics and the modulatory effect of intercurrent illness on dosage.

Norman Gregg and John Cade both had excellent academic records and careers interrupted by gruelling war experience. Gregg had three years on the Western Front in World War I and Cade spent three and a half years as a prisoner of war in Changi in World War II. They returned from war to build clinical practices, and though neither had formal training in science, their acute powers of observation and powerful sense of enquiry led them to critical discoveries.

Though not career scientists, they both thought within the framework of the scientific method. For example, Cade chose bipolar disease to test the idea that metabolites excreted in urine reflect upon a chemical cause of psychiatric disease, because he hypothesised that a rapid change in mood would be associated with a change in urine chemistry. Testing this in guinea pigs, he used lithium urate as a soluble form of the putative urate toxin, controlling the study with lithium carbonate, only to find that both affected guinea pig mood. After working out a dosage by self administration, he proved benefit in manic patients. His article publishing the results has the highest citation index ever recorded for *The Australian Medical Journal*.

In the frame of great Australian discoveries, Gregg and Cade changed their disciplines, and provided models for those who came after. Forty years later, Australians would link the infectious agent *Helicobacter pylori* to peptic ulcer disease, and develop effective eradication therapy enabling proof of causation, again changing the face of medical practice – a legacy of Norman Gregg and John Cade. It is worth noting that seven of the ten most quoted papers in *The Australian Medical Journal* are related to the discoveries of rubella, lithium and helicobacter by practical people whose work was not done in the classical research space.

Geology

At the time of colonisation of Australia, geology was finding its feet. While other areas of the natural sciences helped shape the Enlightenment, geology emerged from it as questions relating to the history of the Earth demanded answers.

The basic concepts of geology were still being formulated when Alexander Berry supplied the first overview of the geology of the east coast of Australia, published in Barron Field's *Memoirs* in 1825. At the start of the 19th century battle lines were drawn between those who believed rocks were formed by crystallisation or precipitation from a primitive ocean (the Neptunists) and those who considered rocks had a volcanic origin followed by cycles of erosion and deposition (the Plutonists).

Late in the 18th century, when Gottlob Werner, a revered Neptunist in Germany, described and classified minerals, he followed the dominant contemporary interest in economic geology and mining rather than theories of earth science. Scientific geology can be thought of as beginning with publications of a leader of the Scottish Enlightenment, James Hutton. In 1788 his "Theory of the Earth" startled the world with the idea of "near limitless time", contrasting starkly with prevailing church teachings. Hutton developed the concept of Plutonism in opposition to Werner's Neptunism, and postulated cycles of erosion and deposition over great time spans. He put forward the idea of uniformitarianism – that processes did not change much with time. From the early 1830s, Charles Lyell's brilliant publications brought these ideas of Hutton together, and presented nomenclature and stratigraphy based on the fossil record – a true junction point between the theory and practice of geology.

The platform built by Lyell was influenced by English and French contributions, none greater than that of William Smith. Smith, a surveyor and engineer, published in 1815 "the map that changed the world" of the stratigraphy of England, Scotland and Wales based on fossil markers and field observations. This map was the first national scale geological map, and became a template for geological mapping around the world.

The self taught European geologists John Lhatsky and Paul Strzelecki used the same principles to produce the first geological maps of the east coast of Australia and the Tasman Peninsula in the mid to late 1830s. They were also influenced by French bio-geologists Cuvier and Brongniart whose time specific localisation of fossil markers of age, identified "secondary" and "tertiary" strata of the Paris Basin. These conclusions were based in part on comparative anatomy. In the 1830s fossils characteristic of older strata (the Palaeozoic era) were described by Murchison and Sedgwick, enabling recognition of strata back to Cambrian times using fossil records. Studies over the following thirty years confirmed that rock strata across Europe maintain a common sequence of genus and species of fossils. The question of course was would that pertain to Australia?

Strzelecki's map, compiled after walking 7,000 miles over five years, was a remarkable attempt by one man in the late 1830s to organise the stratigraphy of eastern Australia into the four epochs (eras) recognised at that time, with detailed comment on the fossil record. However, it contained many errors including confusion of sedimentary and volcanic rocks and generalisations beyond any scientific basis.

Before that, even as the basic ideas of stratigraphy were being developed, there was intense interest in rocks in Australia. Joseph Banks encouraged drilling for coal and collection of rocks especially by Robert Brown. Brown's insightful observations regarding the geographical distribution of Proteaceae in the east and west of Australia was vital support of continental drift, the idea that led to plate tectonics becoming the big unifying idea of geology in the 1960s. Brown's data obtained in the early 1800s preceded Alfred Wegener's continental drift hypothesis of 1912 by more than a century.

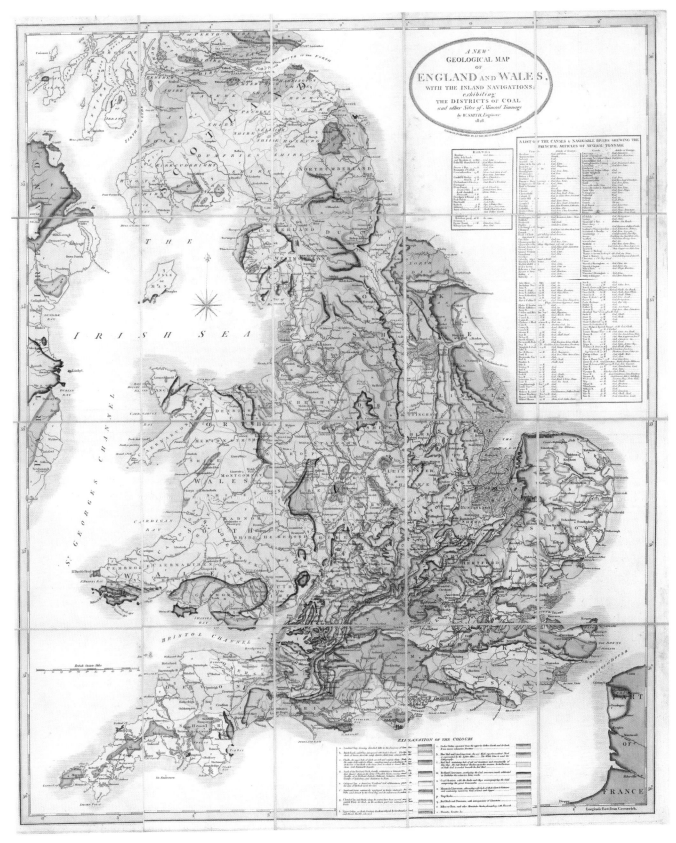

"A New Geological Map of England Wales", W. Smith (J. Carey, 1820)

Geology was a new science still finding its way, just as the geology of Australia was being described and analysed. Smith's map, here in its smaller format, was a foundation document, the first national map of geology based on strata documentation using the fossil record, and thus instrumental in influencing early geologists in Australia, such as Strzelecki, Clarke and Beete Jukes.

Through the early 19th century, explorers such as Oxley, Sturt, Mitchell, Leichhardt and Cunningham collected rocks and made comments on the geomorphology of parts they explored, as did interested amateurs such as Alexander Berry. The discovery of marsupial fossils in the Wellington Caves late in the 1820s produced an extraordinary collection with a list of fifty eight species of fauna, thirty of which were extinct. It attracted the attention of Reverend John Dunmore Lang who focussed on the bones as evidence of the Biblical Deluge as the last catastrophic geological event – a popular theory promoted by the French biologist Cuvier. Sir Thomas Mitchell, the Surveyor General, took a more scientific interest in the Wellington Caves,and the discovery had a major international impact at the time. Lyell stated in his classic *Principles of Geology*, published in 1833, that the fossil bones were evidence for the idea of evolution, and they influenced Charles Darwin's *Law of Succession of Types* published in 1837. Australian marsupial fossils were studied by Sir Richard Owen, curator at the Hunterian Museum of the Royal College of Surgeons. The most important discovery made from bones sent by Mitchell was of the giant marsupial mammoth, *Diprotodon australis*. The watershed, when Australia began to control its scientific destiny in geology, came around 1870. Gerard Krefft, Curator of the Australian Museum, stopped sending key fossils to Owen, retaining them for the Museum and local study, and William Clarke won a battle on the age of the Newcastle coalfields based on field studies, over the British influenced Frederick McCoy, a geologist who based his opinion on English palaeontology, and European Geological stratigraphy.

Reverend William Clarke came to Australia in 1839 and remained as a leader of the scientific community until his death in 1878. He published widely in areas including geology and meteorology, identified most of the basic stratigraphy of New South Wales, contributed to the understanding of goldfield geology and determined the age of the Newcastle coal measures. He resurrected the Philosophical Society and transformed it into the Royal Society of New South Wales, in 1866 as the hub of science in the colony.

Nineteenth century geology in Australia was about finding and exploring economic deposits of minerals – a scientific achievement matching anything done in similarly large areas such as the USA at that time.

The pivotal figure was Edgeworth David who would bookend this period of field survey with his comprehensive map and notes published in 1933. David began his extraordinary career in Australia in 1882 as an assistant geological surveyor under Charles Wilkinson in the New South Wales survey. Geology over the next fifty years in Australia revolved around this man in one way or another. In the first four years he established his credentials as an outstanding field geologist. If proof were needed by the government of the value of geological surveys, it came when David carefully mapped a surface coal seam he discovered in the Newcastle area, and by tracing its subsurface distribution, identified the major South Maitland coalfield.

Geologists in Europe and the United States were moving in new directions that would stimulate interest in late 19th and early 20th century Australia which was developing academic departments in earth sciences. The big questions of international geology were the alignment of Darwin's theory of evolution with the fossil record, especially the apparent absence of fossil evidence of life before the Cambrian period with its well developed metazoan species; the role of glaciation in shaping landform following the Swiss geologist Agassiz's discovery of evidence of a massive ice sheet between the Alps and the Jura Mountains of Pleistocene age; and the development of petrographical geology. Petrographical geology involving the use of thin rock sections for microscopy and chemical analysis dated from attempts by Werner to classify and develop definitions of rocks and minerals. William Nicol in the mid-19th century devised a polarising microscope to examine thin sections so as to analyse structure and identify components such as microfossils – a process refined by Henry Sorby who added the use of spectral analyses to identify chemical constituents, beginning a process of classification based on composition. In the first decade of the 20th century chemical analysis as a basis of classification of igneous rocks initiated new chemical directions in geology.

How did Australian research measure up? One way to judge is to look at the activities in geological research across the colonies at the first Australasian Association for the Advancement of Science meeting in Sydney in 1888, with 800 registrants, about ten per cent of whom attended life science sessions (including geology). The academics Liversidge, Tate and Edgeworth David (who was still working for the New South Wales Government), took a central role in many activities related to geology. Fourteen committees established to continue professional communication included five relevant to geology: "mineral consensus"; geophysical data (including terrestrial magnetism, regional gravity variation and temperature in drill bores); glaciation (including evidence of ice ages from David's studies in Permian coal bearing seams in the Hunter Valley); Antarctic exploration; and "Landforms and Tectonism" – a committee driven by Edgeworth David concerned with a combination of geology and physical geography. Its early multidisciplinary approach was vital in the study of soils and relevant to agriculture and the extension of rural industries and water distribution.

Work monitored by some of these committees would continue for many years dating from their late Colonial beginnings. Another committee was charged with study of igneous and metamorphic rocks. Initially it relied on the enthusiasm of A.W. Hewitt for microscopical petrography. This committee's initial focus on classification gave way to a research interest in alkali-rich igneous rocks found in Western Australia. Study of metamorphic rocks came later, with correlations based on chemical analyses rather than classical (but crude) stratigraphic methods, led by W.R. Browne from Sydney from 1926 and Germaine Joplin around the mid-1930s. By the late 1940s the committee, now known as the Petrological Committee, brought together igneous and metamorphic petrologists.

Development of geological science in late Colonial and early federated Australia progressively centred around academic departments, while government and industry geologists continued the search for minerals, and increasingly for oil and gas deposits (which did not become of economic value until after 1950). In brief overview the AAAS committees actively addressed two of the big questions that concerned northern hemisphere geologists. A third issue, a Precambrian fossil record to support the Darwinian theory of evolution, was a particular Australian success.

Darwin had identified that the major weakness in evidence for his theory of evolution was that the earliest known fossils were biologically advanced metazoan fossils in rocks of the Cambrian era. If his theory was right, Cambrian trilobites must have had simpler ancestors. Australian geologists pursued the search for Precambrian fossils to complete a fossil line demanded by his theory.

These geological surveys evince a dominant academic strand influenced by Edgeworth David who took the inaugural chair of geology at Sydney University. David's impact was profound. Douglas Mawson moved to Adelaide, with Cecil Madigan his student and colleague. David's main objective became a magnum

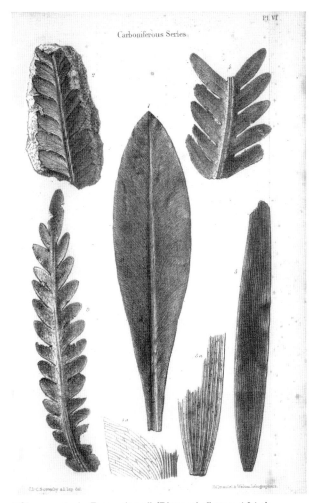

"Glossopteris Browniana" (Plate vi, figure 101a) in *Physical Description of New South Wales and Van Diemen's Land*, P.E. Strzelecki, (1845)

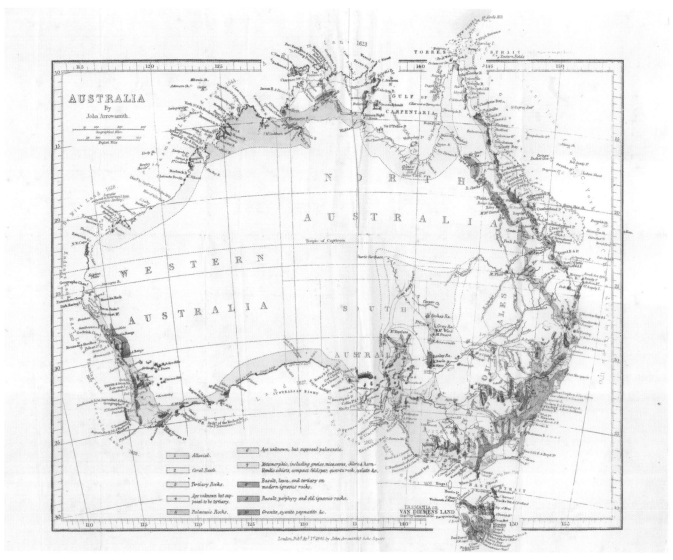

"Australia", J. Arrowsmith in *A Sketch of the Physical Structure of Australia, so Far as it is at Present Known*,
J. Beete Jukes (1850)

Early scientific mapping of the geology of Australia was performed by talented amateurs with scarce and often inaccurate data. A major problem became dating of the Newcastle coal seams, which Strzelecki called Carboniferous based on fossil content as understood from northern hemisphere studies. Clarke identified them correctly as Permian from observation and field studies, which are included in the maps of New South Wales geology (developed in conjunction with the government geologist C.S. Wilkinson).

opus on Australian geology, which was completed by W.R. Browne, another student and his successor at Sydney University. Mawson established his reputation by mapping and analysing the metamorphic series rocks in the Broken Hill area, while Madigan concentrated on desert geology, specially that of the Sturt Desert.

Microbial fossils were found in Precambrian rocks, including examples discovered by Douglas Mawson, as stromatolites or colonies of primitive cyanobacteria embedded in sediment, forming characteristic layers; but the gap between oxygen producing cyanobacteria

and trilobites was too great a leap for evolutionary theory.

Edgeworth David searched Precambrian quartzite at Tea Tree Gully, near Adelaide, and in 1936, claimed to describe "giant annelid and large arthropod" impressions. His peers – including his ex-student and sometime colleague, Douglas Mawson – were sceptical. However a decade later Reginald Sprigg, a student of Mawson's, discovered a metazoan series of arthropod and annelid fossil impressions. When Sprigg described a eurypterid-like animal on the Yorke Peninsula local

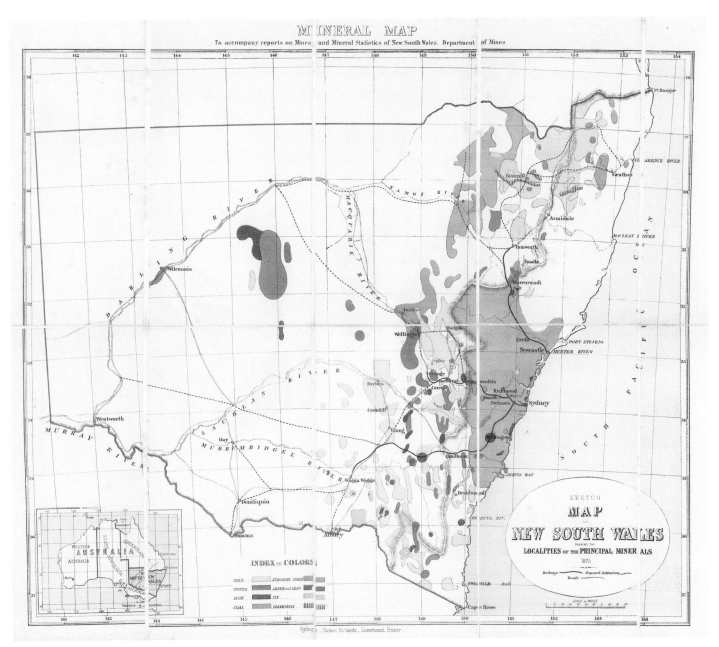

"Sketch Map of New South Wales Shewing the Localities of the Principal Minerals", New South Wales Department of Mines (1875)

palaeontologists again rejected the claims. Persistence was rewarded and in 1947 he found impressions of primitive metazoan organisms of a disc-like shape in quartzite rock, near the Ediacara mine. These were the first fossils of multicellular animals incontestably found in Precambrian rocks, adding a major contribution to the fossil record in relation to the requirement of an evolutionary sequence. The discoveries were published that year in *Nature*. Martin Glaessner, the Professor of Geology at Adelaide, found further arthropod and annelid species, contributing to the recognition that 600 million years ago was a time of metazoan diversification with establishment of the phyla of most invertebrates. Darwin would have been pleased with Australia!

Another pursuit that drew international attention to Australian geology was the search for glaciation and earlier ice ages than were known in the Northern Hemisphere. In 1885 David discovered an extensive area of Permian glaciation in eastern Australia. Many in the 1880s thought the great Pleistocene ice age was unique, so the recognition of an earlier ice age was an important discovery.

Many geologists contributed to defining the two major ice ages in, respectively, the Permian and Carboniferous eras, but none more than Douglas Mawson. His early work in South Australia, especially in the Adelaide System of Precambrian rocks, established an interest in glaciology, which was important in his decision to join Shackleton's 1907 expedition to Antarctica. Mawson's work in South Australia with others in his department did much to define the extent and impact of glaciation.

One hallmark of geologists trained and influenced by Edgeworth David was commitment to, and excellence in, field survey, stratigraphy and structural geology. Firm interest in economic geology was another. Most of the key academics were directly influenced by him. The contributions by Douglas Mawson and W.R. Browne, students of David, were academic foundation stones for geology at the universities of Adelaide and Sydney respectively.

By the mid-20th century a new generation of influential geologists trained or recruited by Mawson was making a mark. Reg Sprigg, Martin Glaessner and Cecil Madigan are discussed below. Another was Harold Reggat (1900–68), an outstanding field geologist who became Director of the Commonwealth Mineral Resource Survey. His versatility enabled critical contributions to the atomic energy programme, the Snowy Mountains Scheme, and in oil, mineral and coal exploration. The close association of Thomas Griffith Taylor (1880–1963) with David at Sydney University led to his participation in Antarctic expeditions and physiographic and economic geology. He was one of few in Earth sciences who gained an 1851 Exhibition scholarship to work in Cambridge. His experience in Antarctica led to interests in meteorology and in landform and the impact of variables such as glaciers and weather. Taylor always focussed on economic outcomes, such as in agriculture.

Walter Woolnaugh (1876–58) was born in New South Wales and completed a degree in geology at Sydney. He accompanied David's expedition to Funafuti Atoll, to test a theory of Darwin's on coral growth. Positions teaching mineralogy and petrology led him to taking the foundation chair of geology in Western Australia (1913–19). As geological advisor to the Commonwealth, he promoted the search for economic oil deposits, as early as 1930, using aerial surveys and photography. He located dome structures in Exmouth Gulf in Western Australia, where oil would be discovered in the 1950s. Like others of his generation he had wide research interests – from localisation of salt deposits and, the lateritic duricrust relevant to soil quality, to geomorphology and the theory of oil formation.

Leading geologists outside Edgeworth David's lineage included Edwin Hills, Keith Bullen and David Thomas. Hills (1906–86) was born and trained in geology in Melbourne and went to Imperial College London on an 1851 Exhibition Scholarship. He became Professor of Geology at Melbourne University in 1944 where he established a laboratory based research unit with a focus on petrography. He was an admired and internationally respected scientist, making a difference in every area of geology in which he worked. He researched fossil fish from the most primitive to recent as parameters of biostratigraphy, as well as acid igneous volcanism, physiography and landform, and structural geology. Despite the breadth of his studies, he achieved recognition in each area.

Keith Bullen (1906–76) was New Zealand born and trained, so it is not surprising his early studies were on the mathematical analyses of seismic waves. They defined his life's work in the physics of the Earth's structure, including wave conduction as evidence supporting a solid inner core to the Earth. He established the most reliable tables in seismology. He became Professor of Applied Mathematics at Sydney University.

The third "non-David" geologist to influence academic geology in Australia was Welsh born David Thomas (1902–78). His life passions of geological surveying, structural geology and graptolite biostratigraphy led to the award of a DSc in 1940 from the University of Melbourne. In senior roles in Victoria and Tasmania he continued his fieldwork with a focus on economic minerals, before becoming Chief Government Geologist in Victoria and Director of Survey. The quality of his survey work and his contribution to the enlargement of the Eildon Dam in 1947 led to him being recognised as Australia's leading structural geologist.

Important Geologists

William Clarke (1798–1878)

Reverend William Clarke is entitled to be known as the Father of Geology in Australia based on his contributions to geological survey and mapping of New South Wales. Clarke was much more than that – he was a catalyst to scientific thinking in mid-19th century Australia, and a coordinating force for science as he resurrected scientific communication in colonial New South Wales through his efforts to re-establish what is now the Royal Society of New South Wales, in 1867. His science was local, but his impact went well beyond boundaries. By his early work surveying "gold areas" around Bathurst, then other areas further south, he established scientific credibility for Australian geology. By creating a communication network, he ensured an Australian voice on the international stage. By standing by his assessment of geological continuity in Newcastle coal seams through careful field observation against opinion based on laboratory work in Cambridge, he broke the yoke of English control, thereby setting up interactive independence of peers on an equal footing. All of this arose over forty years, during which his day job was as rector of St Thomas' Church in North Sydney. His income was marginal, yet he supported his wife and children back in England for education much of this time and coped with the health issues that had encouraged his migration in the first place.

Clarke was not an amateur who made geology a hobby – he was trained by Professor Adam Sedgwick in Cambridge and when he arrived in 1839 had several years of field experience with a bibliography of published papers. He became a trusted government advisor and was offered the chair in geology at Sydney University.

Soon after his arrival he set about geological surveys of the Sydney and Bathurst areas. His work in gold bearing areas got attention but his greatest contribution came with establishing the structure and age of the Sydney Basin. He worked with visiting geologists including James Dana and J. Bette Jukes, and conclusions on conformity and age in the Sydney Basin evolved over years and a period of hostility with the Cambridge School.

His most important publication was his book *Remarks on the Sedimentary Formations of New South Wales*, published by the Government Printer (4th and popular edition, 1878). It is an outstanding review of his (and others') contributions to survey and stratigraphy. It includes comments on other colonies and criticism of work by contempories such as Strzelecki and Jukes. In fact he largely ignored Strzelecki's contribution. Jukes in *A Sketch of the Physical Structure of Australia*, published in 1880, credits Clarke as a collaborator and both had been taught by Sedgwick. But Clarke, for all his outstanding values, was hesitant in sharing success.

Eleanor Grainger's biography, *The Remarkable Rev. Clarke*, is well titled.

Alfred Selwyn (1824–1902)

Discovery of gold in Victoria focussed attention on geology and "mineral survey" and the need for a mineral surveyor. Alfred Selwyn was appointed in May 1852 and immediately started on the geological survey of Victoria – at the extraordinary rate of more than 1,000 square miles a year. Selwyn accepted nothing short of "scientific truth" and within a decade Victoria's geological survey was considered amongst the best in the world. In Australian science Selwyn stood at the junction point between the talented independent (such as William Clarke) and organised academia.

In seventeen years based in Melbourne (1852–69) he completed sixty one maps, and made the first Australian discovery of graptolites in lower Palaeozoic rocks, a reminder of his early experience in Wales. When the Victorian Government closed down its mineral survey Selwyn resigned and moved to the senior position in geology in Canada. The loss to Australia was immense, yet his time in Australia shifted geology from the opportunistic to the systematic, based on the highest of scientific standards. Another legacy of Selwyn's thanks to his rigorous training, was the extraordinary group of talented geologists he trained and influenced, who would promote quality field geology across the colonies – R. Daintree and C. Aplin in Queensland, C. Wilkinson and E. Pitman in New South Wales,

H. Brown in South Australia (and later Western Australia), G. Ulrich in New Zealand, and R. Murray and E. Dunn who remained in Victoria to initiate a recovery of geology thereafter 1872. Perhaps his finest graduate was Robert Etheridge who would become the "go-to" palaeontologist in Australia.

Tannatt William Edgeworth David (1858–1934)

Of Welsh origin with an international presence, Edgeworth David was quintessentially Australian! Certainly, one of the most influential figures in any discipline of science in Australia, revered and respected across the spectrum of society to the extent of being given a state funeral. David's great skill and passion for field survey as the scientific basis for geology, and his commitment to economic geology, stamped his academic philosophy and would become hallmarks of his influence on Australian geology.

Trained in geology in England, he migrated in 1882 and for eight years worked under Charles Wilkinson in the New South Wales Colonial service as an Assistant Geological Surveyor.

He covered much of New South Wales, but like Clarke, made his mark in the colony with surveys of the Newcastle coal measures. He demonstrated the value of field survey by discovering a major coal deposit in the Maitland area of the Hunter region in 1886, from mapping surface markers. In 1890 he took the chair of geology at Sydney University, uncertain that it was a good career move. It was! He got international attention from his studies on the atoll of Funafuti providing evidence in support of Darwin's theory of coral reef formation by progressive growth on a sinking igneous platform. He drilled a bore of 340m and showed organic remains of shallow water marine organisms at depths consistent with a subsiding base. Perhaps his best research was in developing an understanding of glaciation and past ice ages. At the time, most were only familiar with the idea of a single Pleistocene ice age that covered large areas of Europe and North America. In conjunction with a colleague from the Indian Geological Service, R.G. Oldham, David detected evidence of glaciation in the Hunter region of the late Palaeozoic era. He also showed that in Australia Pleistocene glaciation appeared to be

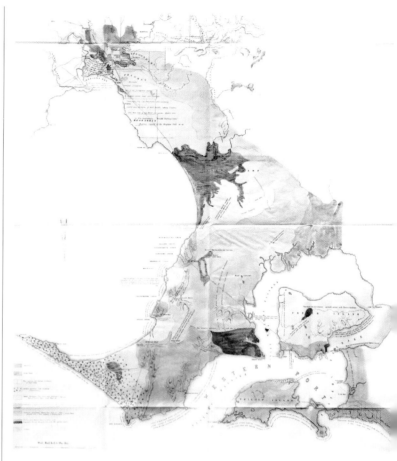

Port Phillip. A. Selwyn (1852)

Selwyn was appointed as the first Geological Surveyor in Victoria in 1852, a time of great interest stimulated by recent gold discoveries. He controlled the colonial geological survey at a high standard for seventeen years. He represented a transition from amateur to professional geology, training competent geologists who would dominate colonial geology in the latter part of the 19th century: R. Daintree and C. Aplin (Queensland); C. Wilkinson and E. Pittman (New South Wales), H. Brown (South Australia and Western Australia); with R. Murray and J. Dunn continuing in Victoria, and R. Etheridge becoming Australia's leading palaeontologist, at the Australian Museum.

restricted to a small area around Mt Kosciuszko – which explains differences between Australia and Europe in soil quality.

His reputation as an authority on glaciation was enhanced when he accompanied Ernest Shackleton to Antarctica in 1907 (at the age of fifty). In addition to studying geology and glaciers there, David led the expedition to locate the site of the South Magnetic Pole, which involved reaching a plateau of 2,700 m altitude and manhauling a sledge 2,200 km with Douglas Mawson and Alistair Mackay—an effort of remarkable endurance and part of the Antarctic legend.

Around 1919, he began his most important project, bringing together the geology of Australia in a single map in the context of an evolving landform over the geological ages. This massive effort was published in 1932 as a map accompanied by explanatory notes, but the task was not complete at his death. David's Sydney University colleague, William Browne, completed it for publication in 1950.

In retirement David continued frenetic activity, which included the search for fossil evidence of Precambrian life, to support Darwin's theory of evolution. His claimed finding of arthropod imprints in Precambrian rocks near Adelaide remains controversial.

Edgeworth David was one of a small group of senior scientists at Sydney University who took on their classical colleagues to establish a firm foundation for the sciences. With his academic friend Archibald Liversidge he was critical to the establishment of the Australasian Association for the Advancement of Science which for many years was the vehicle in Australia for academic communication in geology through a series of committees, most driven by David.

His international connections influenced much activity in academic geology, and his presence established scientific geology in Australia on a sound footing. He had great skill in organising and exploiting new ideas and opportunities—a rare and important talent. Edgeworth David was the only Australian to be awarded the most prestigious award in geology, the Wollaston Medal. Perhaps his greatest contribution was his capacity to inspire loyalty and enthusiasm in others to follow his path—his legacy being the strength of basic geology throughout Australia.

Robert Etheridge (1846–1920)

Born in England and trained in palaeontology by his father and Thomas Huxley, Etheridge spent two periods in Australia, five years as a field geologist recruited to the Victorian survey by Alfred Selwyn at the age of twenty, and later a permanent move as palaeontologist to the New South Wales Government Survey in 1887, recruited by Charles Wilkinson, with whom he had worked in Victoria. Robert Etheridge was an important acquisition for Australian geology, as it had become clear the geological time boundaries in Australian sediments were often blurred because of lack of quality local knowledge of the fossil record. His precise observations and clear notations were included in over 400 publications. His contemporary Edgeworth David applauded his credibility: "the classification and correlations in coal fields, goldfields, arterial water basins, oil fields, and many mineral deposits . . . are based on [his] work."

Among several benchmark publications is *Geology and Palaeontology of Queensland and New Guinea* with Robert Jack. Working on late Palaeozoic fossils in the coal seams

Halysites species, in "A Monograph of the Silurian and Devonian Corals of New South Wales", R. Etheridge (1904)

Etheridge was the "go-to" person in palaeontology in Australia, where his analysis of fossil species helped elucidate important questions related to separation of geological periods, especially when conflict appeared with northern hemisphere data. This plate is from his study of Australian coral fossils in resolving the Silurian-Devonian interface in New South Wales.

of Queensland and New South Wales Etheridge coined the term *Permo-carboniferous* to clarify complex patterns that differ from Northern Hemisphere patterns. His work with Ralph Tate in Adelaide on South Australian Permian fossil beds was equally valuable. In 1893 he was appointed Curator (Director) of the Australian Museum. There, despite ongoing rancour with senior staff, including Charles Anderson who would follow him as Director, he made important changes at every level – including collections, public lectures and new cadetships. He developed an interest in ethnology, promoting that in the Museum as a new section.

As Australia's authority on palaeontology, Etheridge established a line of quality and reliable work on Australian fossils equalled by none.

William Browne (1884–1975)
Edwin Hills (1906–80)

If William Clarke began colonial geological survey and Edgeworth David established geology as a discipline across the colonies, academic and intellectual leadership through the mid-20th century undoubtedly followed from William Browne and Edwin Hills, at the Universities of Sydney and Melbourne. Very different in personality and background, they represented the best of Australian academic geology of their time, and in their own ways were inspirational role models for their students and future leaders in economic and academic geology entering the second half of the 20th century.

Browne, Irish born and a "medical immigrant" (like Halpin, Clarke and David) suffered from tuberculosis. He trained in Sydney in geology. The influence of Edgeworth David inspired loyalty and crafted his pattern of work as part of the "David Dynasty". Well trained in survey geology, Browne contributed new information on the petrology of metamorphic rocks in the Broken Hill region, tackled the Carboniferous/Permian question in the Hunter, and did landmark petrology studies in the Monaro where, by finding graptolites in slate he reframed the age of purported Precambrian metamorphic rocks as Ordovician. Much of Browne's research began with ideas he discussed with Edgeworth David, but in every case, it would become his work and his alone.

This was the case with studies on eruptive rocks of the late Palaeozoic regarding their mineralisation, and studies of glaciation in the Hunter of the Permian-Carboniferous era. His collaborative work over a long period with the State Geological Survey, led by Ernest Andrews, helped with his petrology studies on igneous rocks including the intrusion at Prospect and the lava flows at Kiama. These studies contributed to the development of a synthesis regarding crustal movements and igneous activity in New South Wales. Following studies on early igneous activity, he recognised different types of basalt, which anticipated modern ideas concerning tholeitic and alkali basalts, associated respectively with ocean spreading zones and deeper collisional zones. Browne's study of basaltic batholiths focussed on the time relationships between tectonic activity and deployment of granite bodies.

His most laborious pursuit, for which he is best remembered now, was the fifteen year task of completing Edgeworth David's *The Geology of Australia*. Browne's masterful work transformed the scattered notes David left at his death in 1934, by a huge effort to collate and develop observations across the nation. It was published in 1950 as the statement of the David-Browne era of geology in Australia.

Edwin Hills was a different but equally important geologist – perhaps at an international level better known than Browne, as he related better to his international colleagues. At the age of twenty three with basic geology training from Melbourne University he won an 1851 Exhibition Scholarship which he took at Imperial College to complete a PhD. He returned to an academic career in Melbourne, where he would shine at every level, becoming professor and head of department at thirty-eight. He wrote classic texts on structural geology and physiography. He maintained a classical focus on the importance of the individual and initiative – he was not a team player – but he understood the importance of multidisciplinary approaches to complex problems. While he lent support to a wide and eclectic range of studies, his personal research was along a consistent thematic approach to fossil fish and the biostrata of sediments, acid igneous volcanism, physiography and the evolution of the landscape in northern Australia,

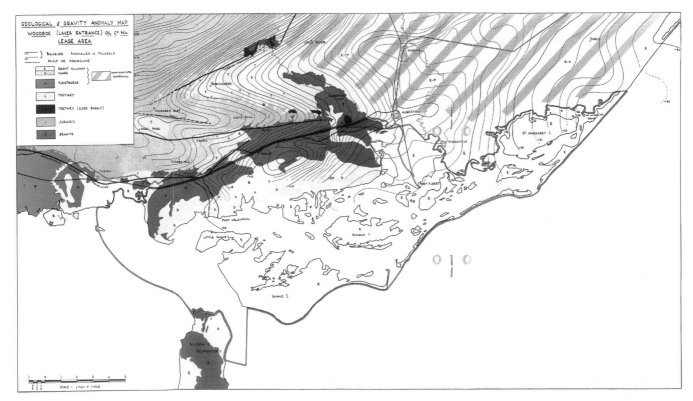

"Geological Gravity Anomaly Map", E. Hills in *Woodside (Lakes Entrance) Prospectus* (1954)

The mould of "practical geology" established by Edgeworth David continued through his students. Edwin Hills took the chair of geology in Melbourne in 1944 having written outstanding texts including *Outlines of Structural Geology*. This map in the prospectus of a company that would drive the oil and gas industry in Australia shows that Hills, like Reg Sprigg, maintained an important role in economic geology.

the Murray Basin and coastal regions. Always he understood the importance of good structural geology. He was constantly aware of the importance of economic geology, from his early work on surveys in the 1940s, to his role representing Australia on international committees in his later years. There he focussed also on arid zones and hydrology. A man for all seasons!

Douglas Mawson (1882–1958)
Cecil Madigan (1889–1947)
Reg Sprigg (1919–94)

Douglas Mawson was a beacon of Australian science. Not the stuff of Nobel Prizes, but a man of focus, belief and total commitment to achieving outcomes denied to others. Best known for heroic activities in the age of heroic activities beyond the shorelines of Antarctica, his career can be difficult to unravel into what was heroic and what was science. Perhaps great chroniclers of Antarctica say it best. Applauding "the scale and

achievements" of Mawson's Australasian Antarctic Expedition (1911–14), J. Gordon Hayes wrote that it was "the most consummate expedition that ever sailed to Antarctica". Sir Archibald Grenfell Price stated that Mawson's was "the greatest contribution to polar science . . . the greatest terrestrial possession". From his mentor and friend Edgeworth David, Mawson inherited the idea of leadership and personal example. Both men shared the eclectic approach to geological science which was a characteristic of the times, respect amongst students and colleagues and across the globe, and a commitment to fieldwork and survey in geological studies.

Summarising his contributions to science is a challenging task. His early influences at Sydney University were Professors David and Liversidge, who encouraged him in early surveys in New Caledonia and chemical analyses of rocks, including work on radioactivity in rocks. That work continued after his move to Adelaide in 1905. In South Australia he began a lifelong interest in glaciation and its geological impact on Precambrian rocks in the Flinders Ranges. His study over several years of the mineralised Precambrian rock series stretching between the Flinders Ranges and the Broken Hill area led to stratigraphic discoveries of the older Willyama series and the newer Proterozoic series, for which he was awarded a DSc. His interest in glaciation led to his appointment as "physicist" on Shackleton's Antarctic Expedition in 1907. It was the beginning of a relationship with Antarctica that has never been equalled. He published work on glaciation and its impact, geomagnetism and geomagnetic phenomena (beginning Australia's interest in the ionosphere and atmospherics), and high quality cartography especially related to his leading role in the party sent from Shackleton's base to locate the South Magnetic Pole with Edgeworth David and A.F. Mackay. Following this epic expedition, David acknowledged Mawson as "the real leader – an Australian Nansen".

Mawson, who always put science first, had little encouragement to continue work in Antarctica from the British explorers Scott and Shackleton, so he launched the epic Australasian Antarctic Expedition in 1911, establishing bases in the "Australian sector", at Macquarie Island and Commonwealth Bay. The prodigious scientific output documented geology, meteorology, geomagnetism, aurora details and biology (including marine science from Captain Davis on the Expedition's ship *Aurora*). Mawson recorded and mapped geomorphology across the Antarctic coast from 90°E to 155°E as well as of Macquarie Island. He introduced radio communication to Antarctica, relayed through a halfway station on Macquarie Island, using technology little more sophisticated than a crystal set. This transmitted meteorological data to Melbourne every day for two years and enabled an accurate longitude to be established at Commonwealth Bay. The Expedition covered 4,000 miles exploring Antarctica, and its published research data filled twenty two volumes released over the next thirty years. Mawson's third and final foray into Antarctica, the British Australian New Zealand Antarctica Research Expedition, sailing over two summer seasons between 1929 and 1931 in the *Discovery*, completed charting of the coast from 43°E to 179°E (Commonwealth Bay to Enderby Land) essentially defining what would become the Australian Antarctic Territory. The Expedition gathered geological and biological observations and collections from Antarctica and Heard Island. The voyages included extensive oceanographic and marine sampling, producing samples for analysis in centres around the world, and published as *The Scientific Reports* in thirteen volumes.

Back in Adelaide after World War I, Mawson continued his inspiring leadership. His research focussed on the Adelaide System of Precambrian rocks – "his territory" – defining glaciation over 1,500 km, and chemically analysing igneous rocks, while keeping up an interest in the biological aspects of the Proterozoic zones, including their stromatolyte and algal content. He continued to identify rare minerals and to look for economic opportunities.

Many important geologists were trained and influenced by Mawson, perpetuating the dynasty begun by Edgeworth David. All understood the importance of fieldwork and careful survey as the basis of geological research. Cecil Madigan was Mawson's closest colleague – as student, partner in Antarctica and academic associate in the University of Adelaide. Madigan delayed a Rhodes Scholarship to accompany Mawson to Antarctica in 1911 and led the party remaining at Commonwealth Bay awaiting Mawson's return from the tragic expedition where he lost X. Mertz and B. Ninnis to vitamin A poisoning and accident respectively.

"Quartzite" in *Geological Investigations in the Broken Hill Area*, D. Mawson (1912)

This section of quartzite from Teatree Gully and the "Adelaide System" contains what David claimed, controversially, to be Pre-Cambrian arthropod fragments, a claim of immense importance in the absence of other clear evidence of evolutionary succession prior to the Cambrian period. David at the time acknowledged that the geological age of the Adelaide System strata was "not finally settled".

In Adelaide Madigan and Mawson divided South Australia for research. Madigan had already completed a geological survey of the Fleurieu Peninsula, and now took on the Central Desert areas.

At the interface between traditional and modern geology, Madigan took advantage of air and water transport. His work with the Air Force across central Australia was the first use of aerial strip photography in Australia for geological survey. The first survey covered an immense area from Broken Hill to the dry salt lakes including Lake Eyre – all over nineteen days. The second mapped land northeast to Birdsville along and over an area he named the Simpson Desert. This raised his interest in inland drainage systems, which he followed with ground surveys including Lake Eyre. He defined the geological structures of MacDonnell Range and associated areas.

His publications included a popular account, *Central Australia*, published in two editions. His interest switched to the study of meteorites and their craters, and the Simpson Desert, an extraordinary sand ridge desert which he crossed with a group of scientists in twenty five days – the last exploration venture in Australia. Again, he wrote a popular account in addition to his scientific papers, *Crossing the Dead Heart*, published in 1946.

If Cecil Madigan was Mawson's most loyal and closest colleague, Reginald Sprigg was "his best ever student". Following graduation, he worked for the South Australian Geological Survey mapping uranium fields at Radium Hill, a site of Mawson's earlier interest, and examined the possible re-working of mines in the Ediacaran Hills. There he found what he thought were jellyfish imprints in quartz. This was the first metazoan fossil identified in Precambrian rocks, a finding of huge importance supporting the application of Darwin's theory of gradual evolution over immense time back to early ages. Sprigg's discovery would lead to the first recognition of a new geological period (the Ediacaran Period) in over a century. Sprigg became a very successful economic geologist, and leader in fuel exploration, finding gas in the Cooper Basin. He was a founder of Santos mining company, and was active in the search for heavy metals including uranium and nickel. His later years combined mineral search and his love of geology with conservation. He acquired an old uranium mining area at Arkaroola in the Flinders Ranges as a site for conservation and ecotourism.

These three great Australian geologists – bridging the stage of field exploration on foot or camel with the modern era of aerial survey and geophysics, were driven by the excitement of discovery and the sheer beauty of what was discovered. It is well caught in Cecil Madigan's introduction to his 1936 book, *Central Australia*, describing his expeditions in the early 1930s in an attempt to convey "the colour and glamour of the vast, lonely, inland region" that "I know and love so well". Madigan expresses what drove geologists of his time with comments about an Antarctic discovery in 1912 in King George V Land, 250 miles from his winter headquarters: "rising from the level white plane of frozen sea the vertical red and colossal pipes of a gigantic organ extending for miles along the coast its hexagonal colour of dolerite stretching skywards unbroken for a thousand feet" seen by "3 pairs of eyes – perhaps never to be seen again".

Frank Debenham (1883–1965)

An Australian geology student of Edgeworth David, Frank Debenham, accepted an invitation from Robert Scott to join his ill fated expedition of 1909–11, during which Scott would reach the South Pole, but not return. Debenham completed most of the expedition's cartography and survey work. He carried out geological surveys in the mountains and glaciers to the west of McMurdo Sound, with skilled use of plane table survey. Returning to Cambridge he became a major apologist for Antarctic research. He was a founder of the Scott Polar Research Institute of which he was Director from 1920 to 1946. This would become a focus for Antarctic research, managed by Debenham with Raymond Priestley and James Wordie. He was appointed Professor of Geography at Cambridge in 1931 and was a prolific author on Antarctica. For many he was the face of modern Antarctic exploration. He was a significant representative Australian involved in scientific documentation of Antarctic research.

Epilogue 1

In the development of scientific enterprise in Australia, natural history remained dominant from its appearance as an Enlightenment science of interest focussing on the exotic flora and fauna of Terra Australis in the Banks era (1770–1820).

While Banks provided the infrastructure and support, it was Robert Brown who established its importance internationally through his comprehensive herbarium using natural taxonomic notation and his insightful comments on geo-botany and biological phenomena. After Brown came a nationalism of necessity, in the climate of the Enlightenment, not dulled by tradition or control – scientists were called scientists and began to call Australia home. They used scientific method to respond to the myriad challenges to survival. In the pursuit of economic natural history, everyone was a scientist. Science clubs for the exclusives became Royal Societies or Linnean Societies, while Mechanics Schools of Arts and newspapers communicated science of the day to an intrigued public, and special groups formed to address unique problems.

Australian ownership of Australian natural history began with Allan Cunningham. Despite a growing sense of national identity the rest of the 19th century was influenced by British science. Paradoxically it was Australian exotics that continued to influence a major connection in British botany with taxonomy. Local taxonomy was championed by Ferdinand von Mueller and Joseph Maiden, European "imports" who became fiercely Australian. Traditional natural history (botany and zoology) had split from geology by the time of settlement as had biomedical science as it related to structure and function (and diseases) of man, while anthropology developed in the era of evolution. The Australian Aboriginal was responsible for leading spectral changes in this new discipline from primitive concepts, through evolutionary then social anthropology. "Economic natural history" dominated throughout Colonial, and well into post-Federation Australia, with agricultural and veterinary science shaped by the realities of survival. This principle influenced the agenda of all scientists of the era.

Biomedical science had its own pace. Though quick to adapt through innovation and initiative, biomedical research was delayed until it was pressed by historic threats of smallpox, bubonic plague and influenza. This section of the natural sciences then accelerated to play a leading role in 20th century science in Australia.

After the Banks era, the two strange bedfellows of imperial Europe and an independent Australia stimulated a pattern of development of natural history similar to that seen in other branches of science: inspired amateur collectors → botanical garden and museum repositories → economic natural history (by individual initiative and departmental programmes) → university departments with ambitious, young British academics → research institutes and despatch of local graduates on 1851 Exhibition Scholarships for experience (including getting a PhD) → return of Australians to run university and government departments, CSIR divisions, and research institutes. In the latter stages, research programmes of internationally competitive scientific platforms are added to economic reality programmes. Key scientists stand at every fusion point.

Natural Philosophy

"The Navigation Sciences"

Natural philosophy was a post–Renaissance classification that included the physical sciences of astronomy and physics, requiring measurement, and mathematics. By the time James Cook brought natural philosophy to Australia, it was firmly established with Isaac Newton's laws of motion and gravity as a unifying principal.

At the time of British expeditions into vast and unknown parts of the Pacific Ocean in the second half of the 18th century, natural science disciplines formed the core of what can be described as Navigation Science, a practical application in the service of the Enlightenment. Objectives included charting a passage (and building knowledge of the world's oceans), accurate location of position in coordinates of latitude and longitude, correction of compass deviation caused by variations in the Earth's magnetic fields and predicting and managing the impact of weather. In advancing these pursuits it was astronomy that got most attention, because compounding star charts was considered the most promising way to estimate time at a reference location (such as Greenwich), and thus by comparison with local time (an easy measurement), to calculate longitude. For navigators in the southern hemisphere, there was a different sky and unknown magnetic fields.

James Cook was the master navigator who brought navigational sciences to Australia. Following colonisation eighteen years later, survey both marine and terrestrial, astronomy, meteorology and the study of terrestrial magnetism were immediately of local value. At the same time they were part of a global effort to ensure safe passage around an Earth that was no longer limited by mythology.

At that time, it was beyond any imagination to anticipate how Australia would provide international leadership in radioastronomy, study of the ionosphere and atmospherics and oceanography. Yet Australia's critical and continuous role in astronomy was heralded over 250 years ago by Captain Cook's expedition – part of a global effort to measure the distance to the Sun. In 1769, by measuring the transit of Venus across the face of the Sun from Tahiti, astronomers expected to determine the scale of the entire solar system, because relative distances between visible bodies within the system were known already. In the 1990s using linked radio-telescopes in Australia in conjunction with optical astronomy and spectroscopy it became possible to measure the minute distortion of a quasar from the periphery of the universe, caused when space is warped by the gravity of an interceding galaxy creating two paths. The measurement of the separation of these paths made possible a determination of the scale of the Universe.

The navigation sciences associated research ideas with navigation challenges. In the West they had been developing since 1535, when Copernicus published his observations of the heliocentric Solar System. Galileo confirmed the findings of Copernicus and developed the idea of a "celestial clock" based on the eclipse of Jupiter's moons as a universal reference point. Kepler built on the celestial observations of Braie to produce a fairly accurate set of tables allowing calculation of position on the Earth's surface. There followed intense discussion and study to find an answer to a leading scientific challenge of the day – the determination of longitude to enable safe and predictable landfall. In 1675 Charles II ordered the construction of an observatory at Greenwich as the English speaking world's first state funded scientific institution to aid navigation, survey and timekeeping. John Flamsteed was appointed astronomical observer.

The Astronomer Royal Nevil Maskelyne in 1763 enabled a reliable determination of "lunar distance" by observing the angle between the Moon and another celestial body and applying a data set available as nautical tables. These constantly updated tables remained in common use until around 1850, when accurate and affordable chronometers became available. Jupiter's "celestial clock" and Maskelyne's method were used by James Cook on the voyage that took him to Australia in 1770.

The concept of terrestrial magnetism was known by the Chinese 2,300 years ago, though the compass to facilitate navigation was not adopted in Western Europe until the 12th century. The importance of correcting for local compass variation was recognised by Spanish and Portuguese navigators facing long voyages across the Atlantic. The Portuguese would miss their halfway stop in the Azores by 400 miles if corrections to compass readings were not made. William Gilbert published the results of his experiments on magnetism in *De Magnete* in 1600, in which he developed the idea that the Earth is a magnet. Edmund Halley made extensive observations on variations in terrestrial magnetism in the Atlantic, published in his *General Chart of the Variation of the Compass* in 1701, with the first use of isogonic lines on a map.

This valuable chart was expanded by Dutch and English navigators including James Cook, with observations made by comparing the needle direction with the Sun's midday shadow.

A resurgence in interest in geomagnetism was initiated early in the 19th century by Alexander von Humboldt, who saw directions not anticipated by earlier scientists. He envisaged prospective studies drawing on multiple simultaneous measurements from a chain of observation points across the British Empire including New South Wales. The French also made important contributions in Australia and the Pacific. The expedition led by Bruni d'Entrecasteaux in 1792 set up a scientific observatory at Recherche Bay in Tasmania as the first organised scientific activity in Australia. Botanists accompanying this expedition, led by Jacques Labillardiere collected 5,000 plants. Important observations made by the expedition's physicist Elisabeth de Rossel on geomagnetism proved that measurements varied with latitude. Additional French observations in the Pacific by Louis de Freycinet in the *Uranie* (1817) and Louis Duperrey in the *La Coquille* (1817–20) added important geomagnetism data sets in response to Humboldt's vision of a world map covering the geomagnetic fields that later would drive the search for the magnetic poles. The South Magnetic Pole was first reached by Edgeworth David, Douglas Mawson and Alistair Mackay in 1909.

"Rossbank Observatory" (1840) in *Voyage to the Southern Seas*, J. Ross (1847)

James Clarke Ross was commissioned by the British Admiralty to explore Antarctica and identify the South Magnetic Pole. He failed to find the Pole as it was well inland, at that time, but he did contribute to the worldwide project initiated by von Humboldt to establish a network of magnetic field observations, by working with John Franklin, Governor of Tasmania, to build the Rossbank Observatory in Hobart.

The Science of Survey

In a vast land, unknown to a settler society bent on occupying it, survey was a foundation science — a counterpoint to celestial mapping and an extension of the navigation sciences.

Survey included observing and recording coastlines in charts, inland discoveries in maps, settlement in cadastral maps, cultural details in topographic maps, and distribution of specific phenomena such as resources, population and physical characteristics in thematic maps. Maps are important documents of scientific endeavour. A major challenge, not met until the 1960s, was completing a national topographic database using rigorous geodetic survey.

Marine survey of harbours, hazards and coastlines around Sydney began immediately after settlement, carried out mainly by William Dawes and John Hunter. There was no detailed charting of any of the Australian coastline, and some areas such as the northeast of Australia were particularly dangerous. Indeed hazards off the west coast had been a preoccupation since the loss of the English ship *Tryall* in 1622, though it took 300 years to establish the location of Tryal Rocks.

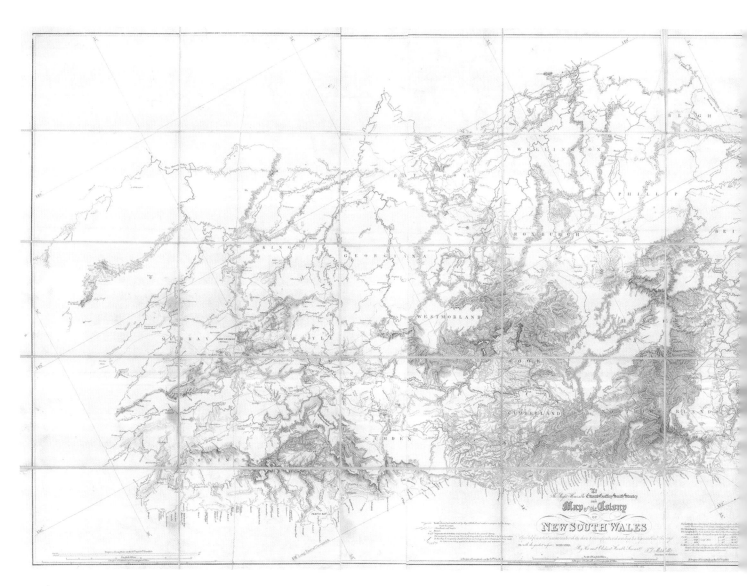

The British Hydrographic Office was established under Alexander Dalrymple in 1795 to chart sealanes relevant to British interests. The Hydrographic Office commissioned navigators, including James Grant, Matthew Flinders and Phillip Parker King to survey the Australian coastline. Charts by Flinders and King made up the "Australian Portfolio" published in 1825. The first detailed sailing directions, by John Septimus Roe, were published in 1830 as part of the *Australian Directory*.

Matthew Flinders' proposal to chart the Australian coastline came at a time when it was not clear whether Australia was a continent or collection of islands, and whether a major river drained a large inland sea. The proposal fell in with the political and scientific objectives of the British Government, the East India Company and Joseph Banks. A feature of these Enlightenment times was the value set on acquisition of knowledge by exploratory expeditions, expressed in the salaries of official scientists in the ships' complements. Each of the scientists accompanying Flinders was paid nearly twice the salary of Flinders, the master of the expedition. The quality of Flinders' observations can be seen in his two maps of the Gulf of Carpentaria, which were used with little change for 150 years. In 1806 he wrote from Mauritius, where he was imprisoned by the French, that having lost the originals, he re-drew his charts from field books using 43 observations of latitude and "871 bearings and angles taken on board [and] 720 with the theodolyte on shore".

Captain Philip Parker King was commissioned to complete Flinders' survey. King was a highly respected member of the colonial community who participated

"Map of the Nineteen Counties", T. Mitchell (1834)

Responding to "the King's Instructions" aimed at defining the limits of settlement for British control of colonisation, Mitchell completed the most ambitious trigonometric survey to that time, with a team of surveyors over a five year period, of land extending approximately 100 miles around Sydney. Base lines were established along Botany Bay and "fixing" location occurred through the Parramatta Observatory, which was "connected" to the baseline using a measured chain.

in the early gatherings of the Australasian Philosophical Society, to which he presented scientific papers on marine survey which were published in Barron Field's *Geographical Memoirs* in 1825. He completed four circumnavigations of Australia and produced fine charts of difficult areas not covered by Flinders – the inner (and safer but slower) passage of the Great Barrier Reef used by ships sailing to England at particular times of the year, and the north west coast from Cape Arnhem to Cape Leeuwin, as well as the south coast of Tasmania. King produced better quality charts than Flinders, some of which would remain in use for over 120 years.

While these early charts were considered adequate for the ships of the day, the advent later of fast clipper ships and steam vessels demanded a higher level of charting accuracy. The Hydrographic Office responded by employing career maritime surveyors, who took control of survey in Australian waters with particular attention to routes along the outer (and faster) passage around the Barrier Reef and through Torres Strait. Between 1837 and 1861 John Wickham, John Lort Stokes, Francis Blackwood, Owen Stanley and Henry Denham conducted continuous survey of coasts and harbours. In 1861 Denham was instructed to negotiate with the colonies to share costs and for them to take a greater role in updating coastal surveys. This was achieved in 1872, a system that continued until 1920 when the Federal Government took sole control of coastal survey.

Land survey well into the 19th century was largely done by men with army or navy experience, commissioned by or working within the Surveyor General's

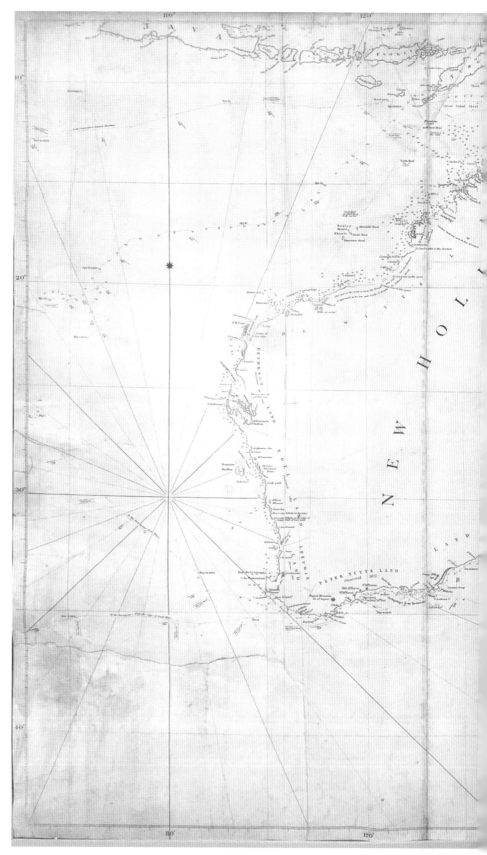

"General Chart of Terra Australis or Australia" in *The Hydrographical Office* (1829).

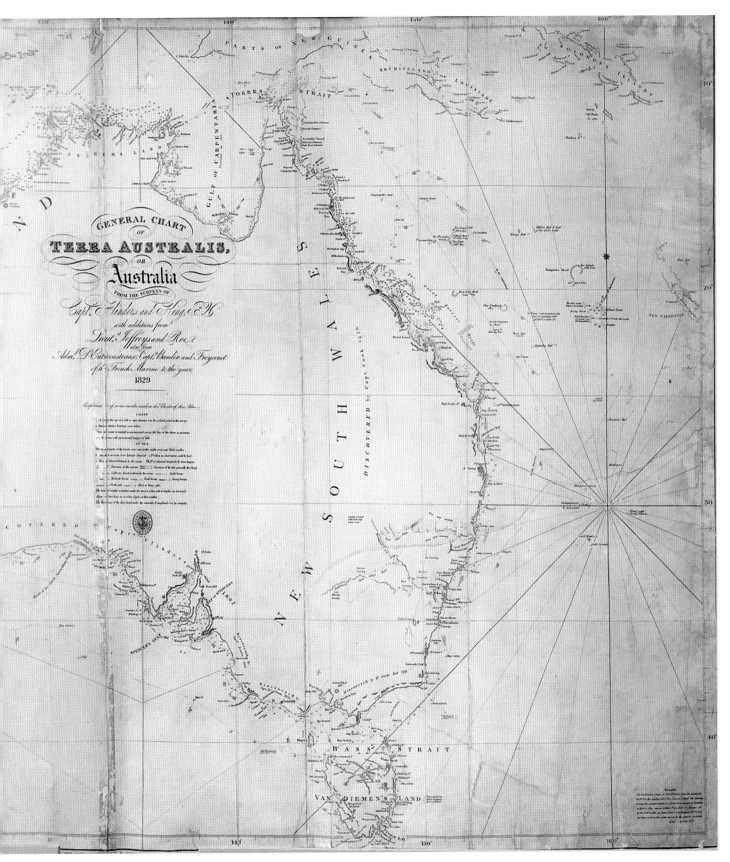

This summary chart of the scientific maritime survey of Australia featured the detailed survey work by Mathew Flinders and P.P. King. Their surveys became the backbone of the "Australian Directory" officially published in 1830. The option of naming "Australia" is noted, as suggested by Flinders on his earlier map published in 1814.

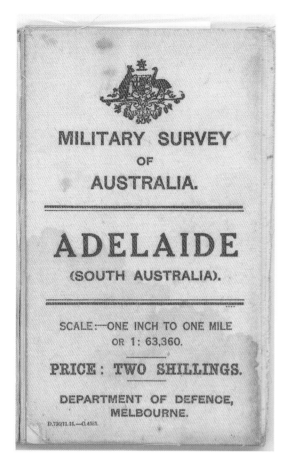

"Military Survey of Australia. Adelaide" (1915), the Government Printer

In 1915 ten military survey maps of regions of Australia were released to the public, one of which was of Adelaide. They were at 1: 63,360 scale, and were the beginning of a national topographic programme of scientific mapping that would become widely recognised as the "one inch to the mile" survey map series, for public and professional use.

department. Those remembered now are largely those who published their journals of exploration, though many high quality maps buried in archives attest the quality of work done by less well known surveyors. The first phase of survey was by explorer surveyors whose maps recording discovery were of sufficient detail to enable road makers, squatters and government officials to follow. This phase of survey is well summarised in the maps published by John Arrowsmith in his *London Atlas*, by John Bartholomew in his *Gazetteer* summarising Australian discovery and growth in the first 100 years of British colonisation and by others.

British colonial discovery and development were Arrowsmith's particular interest. He placed himself politically in the best position to gain new knowledge of inland discovery by being a founding member of the Royal Geographical Society, the official cartographer for Hansard, and closely aligned with great English companies formed to exploit the colonies. Arrowsmith's map "South Eastern Portion of Australia", first published in 1839, went through twelve editions over the next twenty years. A regulation of Governor Darling in 1827 of "the limits of settlement", attempted to confine all settlement in the controlled area 100 miles around Sydney known as the Nineteen Counties. Arrowsmith's maps are a sequence of snapshots of settlement by squatters who ignored those limits. They summarise the way the explorer surveyors met three challenges.

The first challenge was to escape the Cumberland Plain by crossing the Great Dividing Range west of Sydney that had confined the expanding population to the Sydney region for twenty five years. This was achieved when Gregory Blaxland, William Wentworth and William Lawson walked along ridges to avoid sheer cliffs at the end of each valley system. They arrived at Mt Blaxland in twenty one days and viewed the rich pastoral land west of the divide. The science was simple – observe Aboriginal tracks, ignore the prevailing mythology that mountains are all laid out like the Alps in Europe, then test a different idea by following the tracks! Within five years a road across the mountains had been constructed and Bathurst and other settlements established as the vanguard for the Australian future.

The second challenge was to solve the riddle of west-flowing rivers and to test the hypothesis that they flow into a huge inland sea. The Surveyor General John Oxley's attempts at tracing these rivers ended in frustration due to swamps, leaving another navy man Charles Sturt to follow the Murrumbidgee to its junction with the River Murray which has its outlet at Lake Alexandrina adjacent to coastal South Australia. Sturt would later extend his survey to the centre of Australia, without finding anything like an inland sea.

The third challenge was to establish routes of access to the south for which Hamilton Hume and William Hovell were commissioned by the scientist governor

Thomas Brisbane (in 1824), and the north. In the north, botanist Allan Cunningham combined exploration with his search for new plant species, finding passes to the Liverpool Plains in 1823 with encouragement and funding from Brisbane, and to the Darling Downs in 1827. It was these discoveries that encouraged squatters to spread across the colony.

While Arrowsmith's maps accurately presented discovery on a small scale, two sources of larger scale maps would give a contemporary and accurate record of discovery over the complete 19th century. The first was the proceedings of the Royal Geographical Society founded in 1830 "to promote the advancement of geographical science" as a debating society on scientific issues and ideas. The journal published papers by explorers on their recent discoveries including maps as the data source. An excellent record of discoveries in Australia appeared in editions through the middle and late 19th century. The second was a quality contemporary cartographic record in the *Geographische Mitteilungen* directed by August Petermann and published in Gotha, Germany from 1855 to 1945. The maps from these sources are of particular value in recording the data obtained from exploration to meet the next two geographical challenges – to cross Australia from south to north to create a line for the telegraph (and instant communication with the wider world), then to traverse the western deserts to connect the telegraph line with the west coast. The exploration map in Bartholomew's *Gazetteer* best shows that by 1890 areas remaining to be explored were the wide bands between the tracks of Ernest Giles, John Forrest and Peter Warburton in the west, and the area in central Australia around the Simpson Desert (which would be explored and surveyed by Cecil Madigan in the 1930s). John McDouall Stuart, an accomplished surveyor and bushman, completed a south-north crossing in 1862, Construction of the telegraph followed a decade later.

Most of those who forged routes across Australia and discovered new land for pastoral and agricultural use had broader interests in science. They added geology and biology to their agendas, and many important discoveries of Cunningham, Sturt, Leichhardt and Mitchell are amongst the results. For example, both Thomas Mitchell and Ludwig Leichhardt sent Pleistocene megafauna fossils to European experts such as Sir Richard Owen, including giant kangaroos and marsupial lions found in Wellington Caves in Western New South Wales and the first fossil evidence of the giant Diprotodon.

The goal of national mapping was a scientifically accurate survey across Australia, presented as an integrated series of topographical maps. In the event it was not until 1966 that the first homogenous positional data set (the Australian Geodetic Datum) was completed. It required national coordination and standardisation that were never going to happen in Colonial times, when efforts that were begun fell out of register at colonial borders. It took two world wars for government to get the idea, when in 1945 a National Mapping Council was formed.

The importance of a national geodetic survey became clear as we entered a digital age. The shape of the Earth is not regular and is known as a geoid. A geodetic survey scientifically maps the exact shape of the relevant terrain, positioning geographical points in terms of latitude, longitude and elevation. Thus Australia needed to establish a datum relevant to its location on a misshaped earth. Without geodetic survey, locational data of different countries mismatch, often in excess of 100 metres. For global positional systems this is an unacceptable discrepancy.

In the early days of Australian survey there was a focus on cadastral survey without permanent or central reference points. What there was of topographical survey evolved in a nodal sense along expedition tracks. All colonies understood the need for a national trigonometric survey on a defined base line to tie piecemeal surveys and cadasters together. The New South Wales Surveyor General Thomas Mitchell was aware of international thinking about national surveys, especially the Great Trigonometric Survey of India, begun in 1802 by William Lambton and continued by George Everest, but not completed until 1870. Its structure was north-south and east-west gridiron corridors of geodetic survey with corrections to the effect of temperature on measuring chains, making it possible to measure the curvature of the Earth.

When Governor Darling decided to confine all population to within nineteen counties, the Surveyor General Thomas Mitchell saw the opportunity to create a unique topographic survey based on trigonometric measurements. He created two baselines along the shoreline of Botany Bay, each of 832 yards, taking Mt Jellore and Mt Hay as principal triangulation points. To give an absolute location, he tied the baseline to the Parramatta Observatory by Gunther chain (a measured length of chain). The entire programme, completed in five years, consisted of 900 plans. The final composite map was engraved in Sydney and published in London in 1834. It was the greatest scientific achievement in colonial Australia at that time, and the most expansive and comprehensive scientific survey completed outside Europe.

Trigonometric surveys were begun in Tasmania (1833), Western Australia (1829), South Australia (1834) and Victoria (1853). A second geodetic survey was begun in Victoria in 1858 by the Government Astronomer (Robert Lewis Ellery) using meridians of longitude and parallels of latitude, with a focus on definition of crown lands. The construction of a complete map of Tasmania by triangulation between 1833 and 1859 by James Sprent, was an extraordinary effort. His map of the survey published in 1859 stands as one of the major scientific documents produced in colonial Australia.

To print maps in multiple colours was laborious. In 1859, John Osborne, working in the Victorian Department of Lands on geological maps could see the advantages of photography, but he had to find a mechanism for transferring the image to stone or zinc printing plates. To achieve that, Osborne invented a transfer paper of multiple colours, each printed for its own plate. This became a successful system of colour priting in 1860 under Alfred Selwyn in Melbourne, connecting a steam engine to a power press.

After Federation there were many conferences and submissions supported by economic and scientific arguments, usually with little outcome.

The first serious attempt to develop a permanent scientific topographical map department was a military initiative by Colonel W. Bridges in 1909. Initially his section produced a series of maps covering training areas. It then began fieldwork to produce a 1:63,360 (one inch to the mile) topographic map series which combined existing parish maps with primary triangulation and levels for contours measured by aneroid (or provided by the railway department). The first map was of Newcastle, published in 1912. By 1915 twelve maps had been published, covering approximately 10,000 square kilometres.

The Department of the Interior produced nine sheets for the International Map of the World programme between 1926 and 1939. The first sheet of Sydney was essentially an office compilation based on cadastral maps and local reports with spot heights and contours of doubtful value.

In 1924 aerial photography by the RAAF was introduced, and stereoscopic vertical aerial photography in 1927. In 1935 the private sector began aerial survey, though little was achieved before World War II. As with so many early scientific endeavours, it was an individual's initiative and imagination that illustrated what could be achieved by aerial topographic mapping. In this case it was Donald Mackay, a wealthy adventurer, who organised a series of expeditions into central Australia between 1930 and 1936. Government sponsored survey by triangulation on a national grid began in 1951. A primary geodetic network was established across Australia in the eight years between 1957 and 1965. The longstanding problem of a featureless landscape with scarce or unstable reference points was overcome in 1957, when a portable telurometer using radiowaves was trialled from ground level and by air. In 1966 the first scientific homogenous positional data set, the Australian Geodetic Datum, was complete.

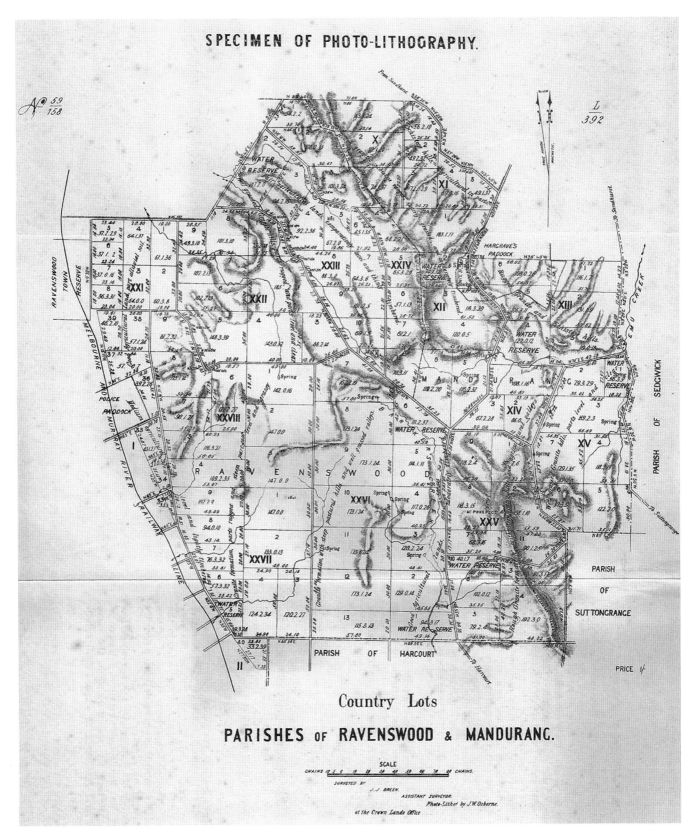

"Specimen of photo-lithography," J. Osborne (1859) in *Transactions of the Philosophical Institute of Victoria* **(1859)**

John Osborne invented the photo-lithographic process of immense value in copying hard images on stone to obtain multiple copies — especially applicable to maps. This would revolutionise copying of maps. The work was done in the Department of Crown Lands, Victoria. It used an intermediary medium and was quickly adapted to zinc plates.

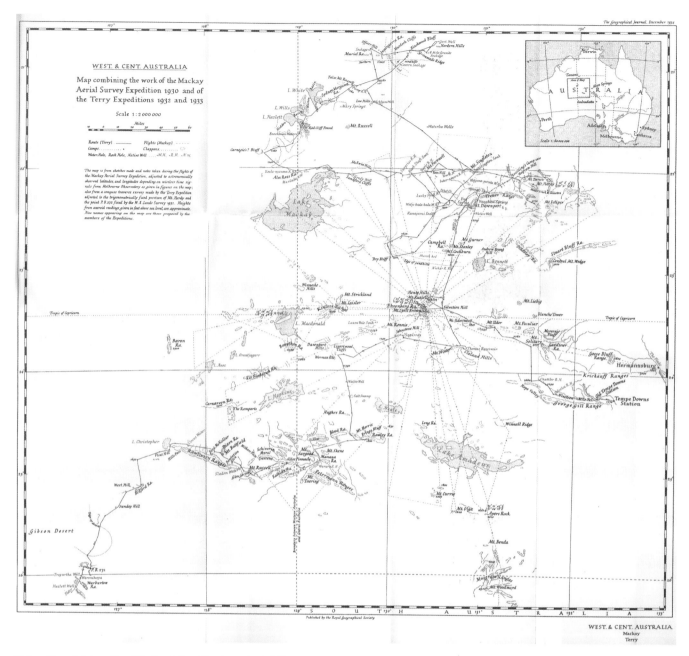

"Map Combining the Work of the Mackay Aerial Survey Expedition 1930 and of the Terry Expeditions 1932 and 1933", D. Mackay (1934)

Donald Mackay was the "last" of Australian explorers, but one of the first to use aerial survey as a tool to address the great distances in central Australia.

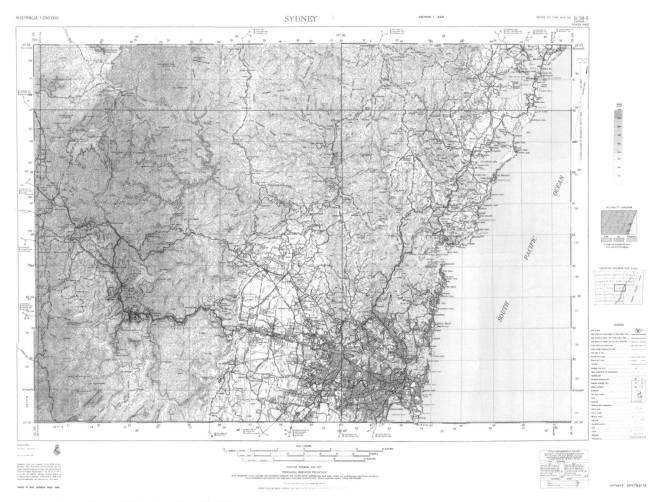

"Sydney" Australia 1:250,000 series (1950)

This was one of the 540 sheets that made up the first completed topographic survey of the continent. The survey for the Sydney map was completed combining photomapping and photogrammetric technology with existing survey maps, by the Royal Australian Survey Corps. Completion of the national survey in 1966 of this first medium scale survey was of particular value for resource assessment.

Important Surveyors

Rear Admiral Phillip Parker King (1791–1856)
Sir Thomas Mitchell (1792–1855)

These two foundation stones of colonial survey were contemporaries, with remarkable similarities in their careers and vast differences in character. Phillip Parker King was of impeccable Australian lineage, seamlessly part of the series of great Australian explorer-surveyors that began with James Cook. He was followed by career marine surveyors of the Hydrographic Office, many of whom he trained when he was captain of the *Beagle* surveying the coast of South America. They included John Stokes, John Wickham, and Owen Stanley. King

was a wealthy, respected and senior member of colonial society, who managed the Australian Agricultural Company through a difficult time in the 1840s, while he also supervised his own extensive pastoral holdings around Sydney. At his core, King was a scientist, recognised amongst his colonial peers as such, taking a leading role in the Philosophical Society of Australia and contributing to Barron Field's *Geographical Memoirs* published in 1825. He belonged to many natural history societies in New South Wales and Europe and was made a Fellow of the Royal Society for his contributions to scientific survey. He indeed was a man for all times.

Thomas Mitchell came from humble Scottish beginnings. Endowed with natural talents he won recognition with the quality of his sketches and plans relating to the Peninsular Wars against Napoleon, and as a result was recommended as an assistant to John Oxley, Surveyor General of New South Wales. Oxley died from tuberculosis, leaving the senior position to Mitchell – which he held, sometimes tenuously, from 1828 until his death twenty seven years later. During this time, he administered badly, fought with every governor, and narrowly missed gaol for debt and death in a duel. He survived through his sheer skills and belligerent persistence, to set the standards for surveying in colonial Australia.

King filled in the coastal survey left by Flinders, completing charts of extraordinary accuracy and completeness. He focussed on the "inner passage" of the Barrier Reef (crucial to safe voyages for much of the year, when it was the best sailing to Asia and Europe), and on the north west of Australia. King's charts made up the core of "The Australian Portfolio" of charts in 1825. He edited sailing directions and contributed broadly to science especially natural history. He summarised his discoveries in a two volume book, illustrated by himself and published in 1826, *Narrative of a Survey of the Intertropical and Western Coasts of Australia: Performed between the Years 1818 and 1822*.

Thomas Mitchell was to land exploration what Phillip Parker King was to marine survey. He explored west of the Great Dividing Range adding to the discoveries of Charles Sturt and John Oxley, especially in the region of the Darling River, then probing north in search of '"the great north river", the mouth of which King had searched for on the coast. Mitchell then went south discovering the rich pastures of western Victoria that he termed Australia Felix. He travelled as far as the Glenelg River.

Mitchell completed precise surveys as he went. On his return, after the seven month expedition into Australia Felix, his progressive survey was only out by an error of 2.8km – an outstanding achievement over the great distance he travelled. Mitchell's lasting memorial was the trigonometric survey of the Nineteen Counties previously described – commissioned by Governor Darling in 1825, in his unsuccessful attempt to establish effective limits of settlement. Robert Dixon and Robert Hoddle,

who would play important roles elsewhere in Australia, were among surveyors working on the 900 surveys done to complete the task. The map of the Nineteen Counties published in 1834 was considered by many (including Mitchell) to be the greatest example of scientific survey of unknown territory to that time, stretching one hundred miles to the north, south and west of Sydney.

Mitchell's four exploratory expeditions were conducted over fifteen years from 1831 to 1846, and recorded in his published journals, *Three Expeditions into the Interior of New South Wales* (1838) and *Expedition into the Interior of Tropical Australia* (1848) illustrated by himself.

Phillip Parker King and Thomas Mitchell were firm supportive friends, respecting each other's scientific talents. They were born and died within months of each other, and shared lifelong interests in exploration and many areas of science, political aspirations and membership of the Legislative Council (from which each resigned because of conflicts). Both had aspirations to commercial success (King through the Australian Agricultural Company, Mitchell by attempting to sell a "boomerang propeller" to the Navy), artistic talents, and large families with sons employed in their business interests (though only Mitchell would send his wife to Scotland to have his mother teach her to be a good housewife). Despite marked personality differences they set similarly high standards for survey, and were role models on how to achieve excellence.

James Sprent (1808–63)

After the start made by Mitchell, trigonometric survey had staggered, incomplete starts in other colonies. The most successful was completed by James Sprent whose twenty five years in surveying for the Tasmanian government resulted in an extraordinary map published in 1859. Sprent arrived in Hobart at twenty two, with a broad and classical education. He established a school but within three years had joined the Surveyor General's Department. Three years later he had completed the trigonometric survey of much of the southeast under the Surveyor General George Frankland. Between 1847 and 1853 he completed much of the remaining survey, often under extreme hardships. This survey included 206 observation stations and cost a total of £20,000.

Major General William Bridges (1861–1915)

The founder of organised topographical mapping in Australia, Major General Bridges, may seem an unlikely Australian scientist of note. A career soldier known for establishing the Royal Military College at Duntroon, and as the creator of the the Australian Imperial Force or AIF, Australia's World War I expeditionary force. His 1st Australian Division was amongst the first ashore

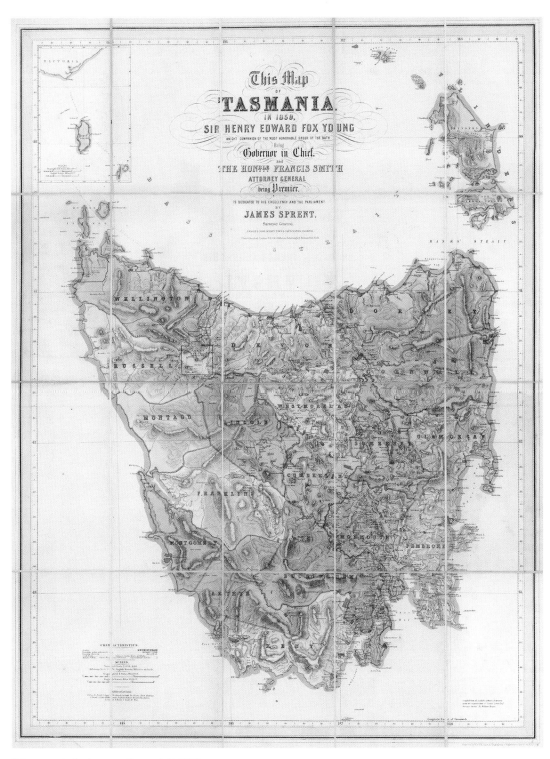

"Tasmania". James Sprent (1859)

This map by Surveyor General James Sprent was the first completed colonial triangulation survey — given the difficulty of much of the terrain, this was considered one of the finest examples of scientific survey at that time.

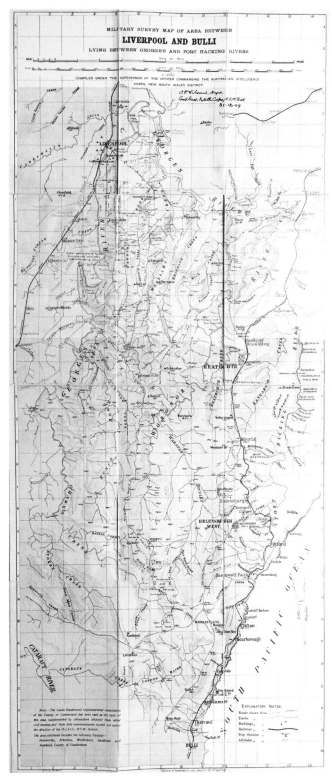

"Military Survey Map of Area Between Liverpool and Bulli",
Australian Intelligence Corps (1909)

One of two maps produced in 1909 by the newly established
Australian Intelligence Corps of training areas, under the
direction of Colonel Bridges. The "one inch to the mile" series
would follow, again on the initiative of Bridges — systematic
scientific topographic mapping had begun.

at Anzac Cove, Gallipoli, where he was fatally injured
by a Turkish sniper. His link to survey and cartography
came in 1905 when he was made Chief of Intelligence
in the Army, and found that very little had been done
to establish a nationwide topographic survey. All that
was available for army training was on a cadastral
base with variable and unreliable topography, yet the
Defence Department was the only structure in the new
Federal system with survey and cartography as a stated
responsibility.

Bridges set up a close communication with the
British armed forces, in particular with Colonel Close,
head of their cartographic division. Close provided
valuable information and support, though he clearly
saw Bridges as a junior and subservient partner. He
recommended two groups of maps: tactical at a scale of
one inch to a mile, and strategic, as a standard smaller
scale topographic map. In the event, Bridges and his
survey department went their own way, printing maps
in Australia controlling both the information and the
process. His immediate need was for maps to support
training and manoeuvres. He gave precise instructions
recognising what was needed, then got it done, noting
that "maps must be done systematically".

In 1909 a few local topographic maps were printed
to support military exercises. As his work continued
Bridges settled on a one inch to the mile series, using
existing triangulation, supplemented as necessary, to
provide a sufficiently dense control pattern for plane
table survey. That is, control points such as trig stations
had to be in site lines for plane table methodology. This
worked well and the first one inch to the mile maps (of
Newcastle and Morna Point) became available in 1913,
printed by the state Department of Lands, in four colours
by the process of photolithography – an invention, by
John Osborne of the Victorian Department of Lands in
1859.

The Science of Astronomy

*Observational astronomy is as old as history itself.
Claudius Ptolemy listed forty eight constellations and over one thousand celestial objects
in his Almagest, a 2nd century mathematical and astronomical treatise.*

However, modern astronomy can be dated from the improved reflecting telescope made by William Herschel (1738–1822). Discovering celestial bodies became the main game of astronomy. Herschel had added significantly to the knowledge of stars, discovering the planet Uranus and its moons and formulating a theory

"Planisphere Australis", M. Abbé de la Caille (1776)

la Caille was arguably the leading French astronomer of his time. His greatest contribution was documentation of the southern skies — poorly understood by those that went before (including Helvetius) — by recording the positions of 9,800 stars in southern skies during a two year stay in Cape Town. This included a number of new constellations represented here using tools and instruments from the arts and sciences, reflecting the impact of the Enlightenment.

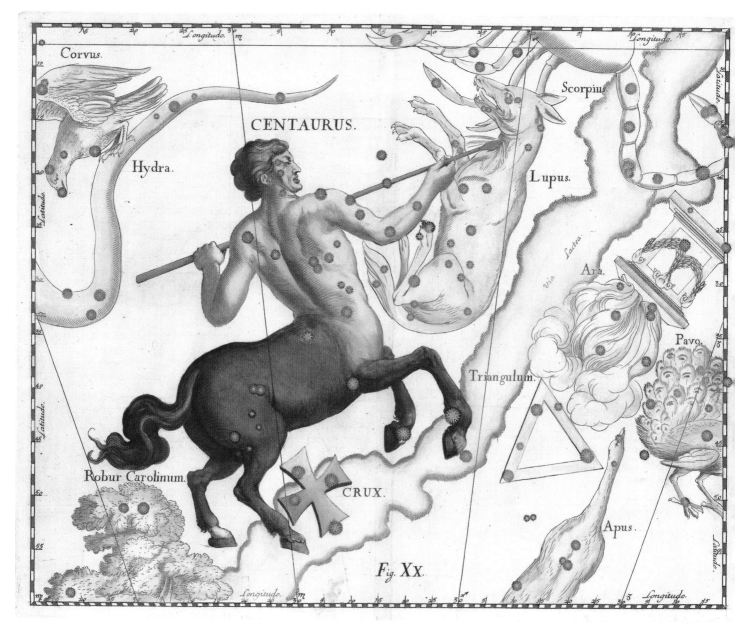

The Southern Cross, J. Helvetius (1687)

Until establishment of national observatories in Paris and London in the 1670s, Helvetius claimed the most advanced observatory, where he produced his own instruments. Considered one of the "Big Four" of the golden age of star map catalogues, his tentatively included limited sightings of the southern sky, and prominently illustrated the Southern Cross, labelled "CRUX".

of stellar evolution based on a study of double stars over time. Herschel added many stars to charts of the northern hemisphere, but little detail existed for the southern skies. Andrea Corsali, an Italian spy working for the Medici family in Florence, travelled to the East Indies with Portuguese merchants, and published the Western World's first chart of the Southern Cross in 1518. This constellation was known to the Portuguese

to be of similar value to navigation in the southern hemisphere, as Polaris indicating the North Celestial Pole. Many navigators added constellations, but none was as valuable as the observations of the Dutch merchant Frederick de Houtman, aided by a spell in a gaol in Sumatra as the Dutch threatened the balance of trade for spices. De Houtman added twelve new constellations which first appeared on the globes of

Petrus Plancius, the official cartographer for the Dutch East Indies Company (the VOC). Edmond Halley added 341 stars to charts of the southern sky following his period on St Helena in 1677. The quantum jump came in 1751–53 from Abbé Nicholas la Caille during a two year residence in Cape Town, when he added 9,800 new stars to the star chart. A true disciple of the Enlightenment, la Caille named his constellations after tools and instruments from the arts and sciences.

The primary aim of James Cook on his first voyage across the Pacific was to observe the 1769 transit of Venus across the Sun on 3 and 4 June at Tahiti as part of a worldwide coordinated effort to estimate the distance of the Sun from the Earth. In the event, blurred contact points caused a variation between measurements by the three observers (Cook, Solander and Green) diminishing the value of the exercise.

The momentum of interest continued with the arrival in New South Wales of the First Fleet in 1788. William Dawes (1762–1836), an officer in the marines, was instructed by the Astronomer Royal to establish an observatory with the particular task of observing the return of a comet expected that year. His many tasks limited the time he could spend in the observatory, and the contribution to astronomy that he could make. The comet was not seen. Dawes left in 1791 and it would be thirty three years before serious astronomical observations were made.

Sir Thomas Brisbane arrived in 1821 to replace Governor Macquarie. It is likely Brisbane's main motivation in taking governorship was to indulge his passion in astronomy, as the southern sky remained largely uncharted. He was far more than an amateur with an interest. He employed two astronomical assistants, Charles Rümker and James Dunlop, with a purpose-built observatory at Parramatta, close to Old Government House.

Rümker's report on his early observations to the Philosophical Society, of which Brisbane was President, was published in Barron Field's *Geographical Memoirs* in 1825. Brisbane struggled in his day job, with friction between himself and the British Secretary of State, at a time when New South Wales society was becoming more complex and more democratic. He departed in

1825 leaving Rümker in charge of his observatory. Rümker was designated Government Astronomer and contributed to the star catalogue, and to the discovery of comets, double stars and other celestial bodies.

Dunlop followed on Rümker's departure in 1831. At Parramatta, Dunlop had discovered 254 double stars and 629 nebulae and star clusters. Following his appointment in 1831 his work diminished, and the observatory closed in 1847. Again, there was a hiatus in Sydney astronomy, that would change in 1858 when a government observatory was constructed overlooking Sydney Harbour.

Thomas Brisbane deserved better than he got. He did a good job as Governor, at a difficult time surrounded by difficult people. His legacy is science. He played a key role in establishing the first scientific think tank in colonial Australia, the Philosophical Society of Australasia. He was a true scientist, as a Fellow of the Royal Society and the recipient of the Gold Medal from the Royal Astronomical Society.

In the second half of the 19th century two strands kept Australian astronomy alive. The first was a collection of government observatories established to provide service to the community. The second, a group of talented amateur astronomers, who developed an international reputation.

Colonial government astronomers in observatories built in all colonies except Queensland were employed to carry out practical roles. They kept meteorological records, were a time reference and a base for trigonometric surveys. They kept magnetic records and developed tide charts. They had little time for research at a critical period for observational astronomy. In Europe astronomy was moving towards astrophysics based on spectrographic analysis. For the first time answers could be pursued to questions about what stars are made of and what were the origins of the universe. Circumstances were not helped in colonial Australia by Sir George Airy, Britain's Astronomer Royal for most of the second half of the 19th century. He imposed scientific imperialism on Australian astronomy. Airy stated that only service aspects such as survey control and meteorological recording were acceptable colonial activities. He appointed astronomers to Sydney and

Adelaide observatories, and wrote their job descriptions.

It is hardly surprising that the Rev. William Scott in Sydney spent his time setting up meteorological stations and recording magnetic field measurements. Scott made one very important observation, not understood at the time. He recorded an observation that correlated sunspots, changes in magnetic field measurements and phenomena such as aurora. This was a prelude to later observations by Australians in Antarctica and studies of the magnetosphere, begun by Douglas Mawson.

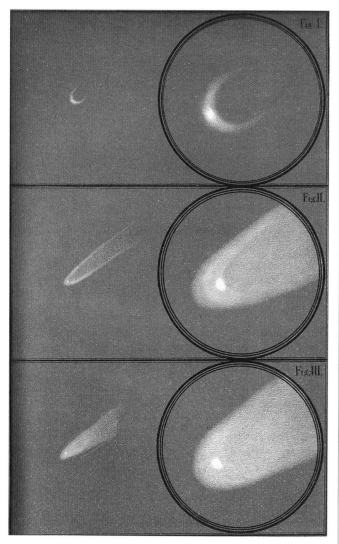

"Donati's Comet", L. Becker, G. Neumayer in *Transactions of the Philosophical Institution of Victoria* (1859)

Amateur astronomers won international respect for Australian astronomy through the mid-19th century. Here the observations of German naturalists and physical scientists Ludwig Becker and Georg Neumayer in Melbourne of the Donati Comet are recorded in the proceedings of the precursor society of the Royal Society of Victoria.

Three government astronomers are worthy of note: Robert Ellery, Henry Russell and William Cooke. Ellery (1827–1908) was the classic amateur of astronomy, arriving from England as a surgeon to make his fortune on the goldfields in 1852, he fell into a position as Director of Victoria's Williamstown Observatory. He would remain head of Melbourne Observatory for forty two years. His memorable act was the acquisition of the Great Melbourne Telescope. Throughout his tenure he maintained quality records of his observations of celestial bodies, and as the quality of work from Parramatta declined, he maintained a reasonable reputation for Australian astronomy. Henry Russell and William Cooke were born and educated in Australia with good local training in optical astronomy. Cooke was trained by Sir Charles Todd (a George Airy recruit) in Adelaide and was appointed as the first government astronomer in Perth.

Henry Russell was the government astronomer who did most to recover a research momentum for astronomy. In New South Wales he inherited a service department with little commitment to discovery and began a research programme recording double stars and transit observations. He acquired new equipment with which he revived John Herschel's earlier studies in Cape Town on displacement of double stars (a double star is two stars that may be in mutual orbit gravitationally bound to each other, or simply in line of sight alignment). He initiated involvement by Australian astronomers in the Carte du Ciel project – a world first attempt to record and map all celestial bodies using the new technology of dry plate photography. This would include observations in Sydney, Melbourne, Adelaide and Perth. This huge international programme was an all consuming project that gave little time for research in much else.

Two other individuals linked to colonial observatories deserve mention: Sir John Franklin and Georg von Neumayer. Both made their reputation in relation to polar exploration and research, but also played important roles in Australia.

Franklin, who had sailed with Matthew Flinders, was Lieutenant-Governor of Van Diemen's Land from 1837 to 1843. Hobart of 1837 was a backwater and half the colony's population of 42,000 were convicts. Driven to change this, he began the Tasmanian Society of Natural History and the Royal Society of Van Diemen's Land. His wife was responsible for the first museum in Australia. His opportunity to become involved in serious research came with the visit of James Ross, en route to Antarctica. Ross, a fellow Arctic traveller, had located the North Magnetic Pole on the Boothia Peninsula in Canada and then conducted the magnetic survey of England. By this time von Humboldt had begun his movement for integrated earth studies and encouraged Carl Gauss to become interested in terrestrial magnetism. Gauss had developed a mechanism for calculating the direction and the strength of the Earth's magnetic field, making it a realistic endeavour to have a global network of observation posts.

Ross had been commissioned to establish several such posts, especially in Hobart because of its southerly location, and to seek the remaining Holy Grail – the South Magnetic Pole. He could not reach the Pole which lay in a still inaccessible inland Antarctic site. Ross and Franklin established the Rossbank Observatory in the grounds of Government House, where continuous observations of magnetism and weather were made from 1840 until it closed in 1854. The data collected became background information to solar astronomy in Australia. Much later Hobart became the site for low frequency radioastronomy studies, because of its southerly position.

Von Neumayer came to Australia in 1851 to find gold but found science and a lifetime love of geophysical and oceanographic studies, and a passion for Antarctica. During ten years in Australia he completed a magnetic study of Victoria. He returned to Germany to become a leading figure in geophysics.

The second strand was a group of talented amateurs who brought more international attention than did the professionals. They were scattered across the colonies, and not restricted by service requirements. Several had high quality instruments and published in local and international journals. They also reported findings in local newspapers which kept the excitement of discovery before the public.

Francis Abbott (1799–1883) was a watch and clock maker who arrived in Hobart as a convict for five years before beginning a successful business and re-starting his interest in astronomy. Initially he used Rossbank, then his own observatory. For many years he published meteorological findings, and made many novel observations, publishing locally and in Europe. He was an active fellow in the Royal Astronomical Society which published many of his observations.

John Tebbutt was an internationally respected astronomer in New South Wales who turned down offers to become a professional astronomer. He preferred being a farmer, and independence. He was a prolific publisher of new observations in the local press, especially the Proceedings of the Royal Society of New South Wales and in international speciality journals. His discovery of the Great Comet of 1861 changed the way Australian astronomy was seen by his international peers. He followed that discovery with many important observations on planetary phenomena and double and variable stars. He produced regular bulletins on meteorology and tidal information until 1907.

The observational skills of amateur astronomers have always had an important place in Australian astronomy. However, the high cost of equipment and the complexity of technology, mathematics and computing, restricted the level of their contribution. With Federation, it was time for a national government to step in and support Australian initiative. But first, we need to look at where astronomy outside Australia was going. Conceptually the situation was similar to botany. While German botanists had focussed on laboratory studies of structure and function, Australia was stuck in a British-style holding pattern with herbaria and phylogeny.

The second half of the 19th century saw the beginnings of astrophysics – a transforming phase when an aged "observational astronomy" switched gear to embrace the promise of astrophysics – to analyse the very dynamics and origins of a turbulent universe. The breakthrough was an observation by G. Kirchoff in Germany, that the solar visual light spectrum included lines characteristic of certain elements. This was immediately recognised as a game changer – here was a tool to determine the composition of stars and their surrounds. Only the Sun produced light strong enough for spectroscope analysis of that time. Solar eclipses gave opportunity to isolate a

"The Great Melbourne Telescope" from *Explorers of the Southern Sky*, R. Haynes, R. Haynes, D. Malin, R. McGee (1996).

Government involvement with astronomy began with the appointment of Colonial Astronomers and the building of observatories. All colonies would share in this endeavour, though the contributions at Sydney by Henry Russell and Melbourne by Robert Ellery led the way. Service requirements became onerous for all Colonial Astronomers — meteorological recordings were part of routine expectations, leaving little time for research. Hence the value of the amateur astronomer.

light source. By the late 1860s the yellow line of helium was discovered. An unparalleled period of discovery had begun, combining the discipline of astronomy with spectroscopy, chemistry and photometry. Albert Le Suer in Melbourne, Henry Russell in Sydney (and some amateurs such as Francis Abbott in Hobart) began limited spectroscopic studies – too late and too little to put Australia in the lead. Catch-up in Australia had to await a new way of thinking where costs and complexities were of a different order. Physics and mathematics would be fundamental to advances in astronomy in all its forms.

The fundamental physics of new age astronomy of all forms in the 20th century was recognition that charged subatomic particles in matter emit electromagnetic radiation. The spectrum of electromagnetic radiation ranges from low frequency long wavelength radiowaves, through infra-red, visible, ultraviolet to x-rays and gamma rays with the shortest wavelength. Different conditions create wave forms of different wavelength. As this was recognised, radioastronomy then infra-red astronomy developed, each revealing complementary visions of the universe. Australia would become a world leader in radioastronomy immediately after World War II – a position it maintains as a principal in the Square Kilometre Array project at Murchison in Western Australia, begun in 2012. This ambitious project combines a square kilometre of radio receptors tuned to detect the most distant long wavelength signals of a primordial and expanding universe, handling more than 5 terabytes of data per second at speeds 10,000 times those of current computing facilities. By recording data from first light to the first half billion years following the Big Bang, an evolutionary map of the Universe can be constructed and evidence for extra-terrestrial life sought.

To account for this dominant position in astronomy, the contributions of the Mt Stromlo Telescope, of physics to the academic development of astronomy in Australia, and of the CSIR ground floor research into radioastronomy after the War need to be understood.

The period between identification of electromagnetic radiation by James Maxwell in 1865, and Ernest Rutherford's model of the atomic nucleus in 1911 saw the discovery of x-rays, electrons and radioactivity, as well as development of quantum mechanics to describe the physics of atoms and subatomic particles, and the General Theory of Relativity describing gravity as a geometric property of space and time. These developments thrust physics and astronomy to the forefront of 20th century science. The evolution of astrophysics, aimed at the when and how of the universe, changed astronomy to a mathematics based physical science, founded on those extraordinary discoveries.

Australia played no part in this revolution except for the contribution of William Bragg and his son, Lawrence. In 1895, Wilhelm Röntgen and Guglielmo Marconi demonstrated, respectively, x-rays and radiowaves, at opposite ends of the electromagnetic spectrum. Both these wave forms interested William Bragg who had taken the chair of mathematics and physics at Adelaide University. He arrived as a pure mathematician, and left Adelaide after twenty two years as an internationally recognised physicist with an interest in electromagnetic waves. His Australian born and educated son, Lawrence, would share the Nobel Prize with his father in 1915 for their studies on x-ray diffraction to identify crystal structure. Bragg's work in Adelaide began within months of the discovery of x-rays by Wilhelm Röntgen in 1895, with experimental production followed by ionisation of gases and diffraction of x-rays by matter. Bragg had a powerful influence in the promotion of physics locally and across Australia, with students including J.P.V. Madsen who instigated the wartime radiophysics laboratory developing radar for CSIR, and who would become the first Head of Astronomy in Western Australia.

Modern astrophysics had to wait the opening of the observatory at Mt Stromlo near Canberra. This came about through the vision and energy of Walter Duffield from Adelaide, who had gone to England to do postgraduate work with William Bragg. He recognised the importance of continuous monitoring of solar flares in the context of terrestrial magnetism and saw the opportunity and value of developing a solar laboratory in the western Pacific region. Astutely, he promoted his idea in Canberra, avoiding state jealousies and taking advantage of a political need for kudos within the new national capital. Duffield was appointed Director in 1924 and the observatory opened in 1926.

Early studies connected solar events with ionosphere disturbance and radio fadeouts, an observation of value in timing communications in World War II. Other studies included the ionosphere and conductivity, cosmic rays, and analysis of Fraunhofer dark lines in the solar spectrum. Clabon Allen, a physicist from Western Australia produced a photometric atlas of the solar spectrum. Duffield established an important connection with the new postgraduate Australian National University. Joint appointments and graduate studies began – the first PhDs and DScs awarded by the University were in astronomy. Formal amalgamation came in 1957, and a second observatory was built at Siding Springs in the Warrumbungle National Park, away from the light distraction of Canberra.

The second director was Richard Woolley (1939–55) – an English astronomer trained at Greenwich Observatory. He switched research emphasis from solar to stellar studies. Woolley established the Australian Radio Propagation Committee; Clabon Allen, who was asked to investigate fade out, found that it resulted from failure of the ionosphere to reflect radio waves in the expected fashion – hence. Allen elucidated characteristics of so called M regions, and showed major storms were associated with sun flares and sunspots, with a day's delay. His studies on recurrent storms and particle size produced the first ideas relating to solar wind.

He also found that Fraunhofer dark lines in the solar spectrum are due to interplanetary dust. Allen became the link man between Mt Stromlo and Joseph Pawsey, a radiophysicist who had been working on radar research during World War II. Together they began measuring radio emissions from the Milky Way galaxy.

As spectroscopy in the latter part of the 19th century was the technology that gave birth to astrophysics and led to recognition of chemical space and a dynamic and turbulent universe, so the recognition that light is but one component of the electromagnetic spectrum enabled observations that were determined by the particular wavelength being used to make the observation. There was nothing special about visible light! The two new "wavelength optics" were radioastronomy and infra-red astronomy. (Infra-red astronomy developed after the period reviewed in this discussion).

Radioastronomy grew out of secret wartime development of radar, based on detection of radio interference. A similar radio interference had been noted in 1932 by Karl Jansky working in the USA and traced to a source beyond the solar system with wavelengths from the millimetre end of the electromagnetic spectrum to wavelengths of hundreds of metres. Before World War II the CSIR had established the Australian Radio Research Organisation with the goal of improving transmission over vast inland distances of rural Australia. During the War England requested help developing radio direction finding or RADAR, using the distances of Australian geography. After the War, David Rivett as CEO of the CSIR, saw value in keeping his commitment to the inclusion of both basic and applied research. Joseph Pawsey led the radioastronomy group to remarkable early success. The first seven years (1945–52) witnessed a crescendo of significant discoveries, putting Australia at the leading edge of this new and exciting form of astronomy. It resulted in David Martyn (the first Australian scientist to work with British research into radar in 1939) coordinating an international scientific meeting in Australia – the 10th General Assembly of the International Scientific Radio Union. This was a beginning for Australian science, which over the next fifty years would host international meetings on a rotary basis in nearly all disciplines of science.

Pawsey's group, with David Martyn who had moved to Mt Stromlo, published data relating variation in wavelength to solar temperature which was recognised as the real beginning of radioastronomy. By the clever use of reflected waves off the sea, interference between direct and reflected recordings opened up the concept of creating a map of the radio picture of the Sun. The idea of multiple site interferometer spacing, developed by Chris Christiansen, produced an enormous data load that could not be handled by the limited computing on hand – but was available to competitors in Cambridge who were later awarded a Nobel Prize for their work. The fieldwork for this study used multiple stations scattered around Sydney. Christiansen set up a series of antennae which required running 200m every couple of minutes to maintain direction of the antennae. He identified "radioplages" or areas of intense radio emissions, mapping two dimensional distribution of radio emissions. He modified his two lines of antennas to follow the Sun's movement, changing the angle and thus enabling the construction of the two dimensional map of the radiosun. It was the first use of the Earth's rotation to produce a "synthesis map" of this type. The CSIR's mistake was not realising the importance of developing electronic computing to analyse huge databases such as Christiansen's. This growth phase between 1947 and 1960 involved many talented physicists, including John Bolton (cosmic ray studies), Paul Wild (solar physics), and Bernard Mills (radio-linked interferometry). Equipment was upgraded to improve angular resolution to obtain finer detail. After 1960 there followed a period of "technique-orientated research" which included building the Parkes radio telescope (1959–61), the first in the Southern Hemisphere built as a large moveable dish.

Above: Dover Heights Field Station, 1951, from *Explorers of the Southern Sky*, R. Haynes, R. Haynes, D. Malin, R. McGee (1996)

Left: The Potts Hill Field Station, from *Explorers of the Southern Sky*

These illustrations record the beginning of investigative radioastronomy, with the innovative interferometer at Dover Heights measuring waves reflected from the ocean, and the Potts Hill site near Sydney where early observations on the Sun and stars were completed using arrays.

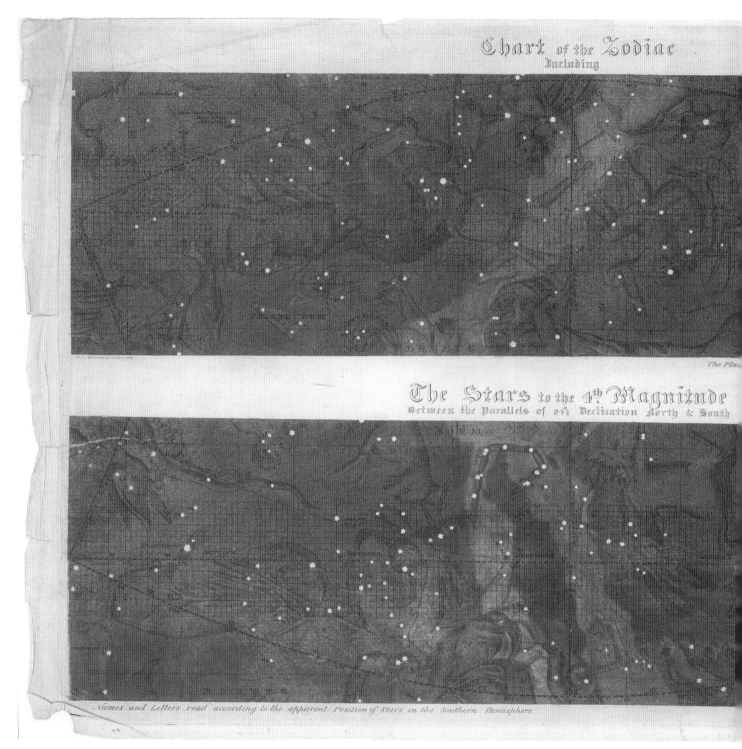

"Chart of the Zodiac Including the Stars of the 4th Magnitude between the Parallels of 24°½ North and South", T. Mitchell (1831)

This chart presents data collected from the Parramatta Observatory established by Governor Brisbane in the early 1820s. The observations were largely made by James Dunlop, who accompanied Brisbane to Sydney to establish an observatory. A catalogue of 7,385 sitings, including those used to complete this map, was published by William Richardson in 1835.

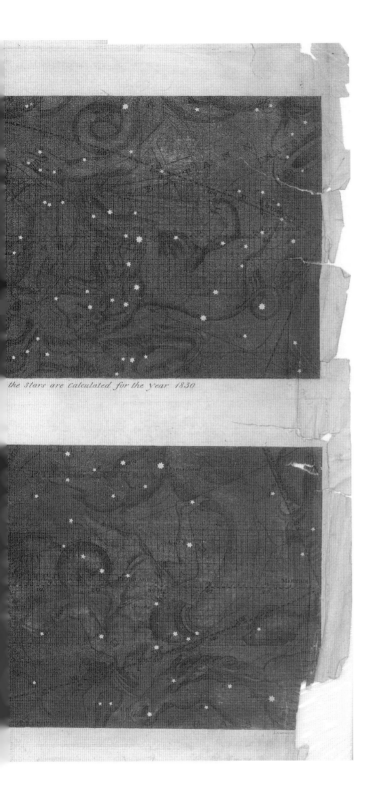

the Stars are calculated for the year 1830

Important Astronomers

Thomas Brisbane (1773–1860)
Carl Rümker (1788–1862)

The discovery of Encke's Comet on 2 June 1822 by Carl Rümker, at the Parramatta Observatory built by Governor Thomas Brisbane, marked the moment the world paid attention to Australian science. This was only the second comet to reappear in a mathematically predicted way (Halley's Comet being the first) and was anticipated throughout the scientific world. Its direction confirmed the value of the physical principles and mathematical formulae underpinning views of cosmology. Encke's Comet was not observed in the Northern Hemisphere nor in South Africa, and Rümker's observations brought him international acclaim. His "Astronomical Observations Made at the Observatory at Parramatta" would be published as a supplement to *Philosophical Transactions* of the Royal Society in 1829.

Thomas Brisbane at forty eight years was already a recognised astronomer and Fellow of the Royal Society when he replaced Governor Macquarie in 1820. There is little doubt that the main attraction to him of becoming Governor of New South Wales was the opportunity to build and run a private observatory, to explore the southern skies which in the past had only fleetingly been examined.

Brisbane brought with him Rümker, and James Dunlop who had no real skills in astronomy, but technically was adept and maintained the instruments Brisbane imported. Rümker was a self taught astronomer from Germany with considerable mathematical skills. The energetic trio immediately began recording the position of stars between the zenith (the point on the celestial sphere directly above the observer) and the south celestial pole – a protracted task eventually leading to a publication of 7,385 stars in the Parramatta Catalogue. The Parramatta Observatory also provided useful meteorological and time reference data for the colony and Rümker and Dunlop at different times continued observations. However, the observatory never recovered from Brisbane's early recall at the end of 1824. Rümker benefitted in the longer term with recognition by the

Royal Society and appointment to the prestigious position of Astronomer at Hamburg. He was an example of a long line of young unknown but intelligent and enthusiastic European scientists who having established credentials in colonial Australia, returned to Europe and a stellar career – a pattern beginning with Robert Brown that would develop momentum in both science-driven business and academia.

Brisbane was never given credit for many excellent administrative tasks. He began a process of bringing squatters under government control, set up agricultural training schemes, introduced more humane approaches to convicts and attempted to maximise their skills and experiences to make a worthwhile contribution to the colony. He directed attention to the documentation of survey of lands and in his own words, was committed to getting "the colony on to its own resources". But in science he did achieve recognition. He was chosen as president of the first scientific organisation in Australia, the Philosophical Society of Australia and contributed, with Rümker, to the papers in Barron Field's *Geographical Memoirs* published in 1825, specifically on local astronomy and meteorology. Thomas Brisbane gave a foundation and energy to the new "Australian Science". He was responsible for the first Australian based research to win a high level of recognition by the great British astronomer of the times John Herschel (in relation to the detection of Encke's Comet). Herschel said it was "a trophy at once of the certainty of our theories and the progress of our civilisation".

Henry Russell (1836–1907)

The first Australia born Colonial astronomer held his position from 1870 for more than thirty years during which he won for Australian astronomy continuity of international attention, and helped establish a collegiate basis for Australian science. Restricted by regulations from Britain's Astronomer Royal, who saw Colonial astronomers as functionaries, Russell's enormous energy and innovation enabled him to simplify his tasks through delegation and concentrate on astronomy. Although he had to carry out meteorological and hydrological studies, he managed to make significant advances in three areas of astronomy.

First, he began a long term project in double stars. He was able to contrast current data with those documented by Herschel forty years earlier in South Africa, enabling him to calculate orbital information which led to a calculation of star mass – information of great future value as astrophysical studies probed inner characteristics of stars.

Second, he took advantage of every opportunity to gain data from transits; he organised a massive effort for the eclipse of 1874, with observation stations scattered around New South Wales which produced 220 photographic plates supplying important data. Russell's photos and drawings were recognised internationally as the most complete of all recordings across a world focussed on this eclipse.

Third, he led Australian observation in the great world astronomical venture, the Carte du Ciel, organised from Paris in the mid-1880s. This project aimed at photographing the sky in great detail using dry gelatin plate photography introduced in 1880. There were nineteen observatories around the world, in a project aimed at producing 22,000 plates. Problems in plate analysis, and with most units not seeing the project as particularly valuable, prevented it having the impact expected, and for Australia, it delayed other research opportunities by sequestering scarce resources.

Russell was also interested in meteorology and had considerable abilities in construction of delicate equipment. He studied water run off, artesian collections and river flow, and even the effect of bushfires on rainfall. He worked at the time when research was becoming organised, and made major contributions to professional societies. He was involved with a core group centred around Archibald Liversidge, building science in the University of Sydney, the Royal Society of New South Wales, and the Australasian Association for the Advancement of Science. Russell was President of the inaugural meeting of the AAAS in Sydney.

Georg von Neumayer (1826–1909)

One of the most remarkable polymath scientists of the 19th century, with Baron Friedrich von Humboldt as role model and support, von Neumayer spent seven years in Victoria establishing baseline physical observations from his privately funded Flagstaff Observatory in Melbourne. He completed a magnetic survey of Victoria between 1858 and 1864, travelling 11,000 miles and setting up 230 magnetic stations, inspired by his mentor's desire to complete a world magnetic survey. Melbourne was chosen largely because of its southerly position. The exploration and scientific study of Antarctica became Von Neumayer's great passion, and he saw his studies in Australia as a precursor to Antarctic science. He became closely associated with the "Heroic Era" of Antarctic exploration, promoting German involvement and chairing of the International Polar Commission of 1879. He also chaired the German South Polar Commission that coordinated the German national expedition led by Erich von Drygalski in 1903 to the area immediately south of Australia.

In Melbourne, first at the Flagstaff Observatory and then in the Botanic Gardens, von Neumayer established a uniform system for meteorological observations in Victoria, and gradually became recognised for his contributions. In his report of 1859/60, "Magnetical, Nautical and Meterological Observations" at Flagstaff Hill, he stressed his belief "that the rapid and enlightened progress of the Colony of Victoria would ensure for physical science a powerful support in the Southern Hemisphere". He was deeply involved in the scientific life which was gathering strength in Victoria in the aftermath of the goldrush, mainly through his involvement in the Royal Society of Victoria.

John Tebbutt (1834–1916)

Sydney born John Tebbutt was the star of a large group of amateur astronomers examining southern skies in the great European tradition. He recorded the important celestial events as they occurred, especially between 1874 and 1887: two transits of Venus 1874 and 1882, two transits of Mercury 1878 and 1881 and the four great comets of 1880, 1881, 1882 and 1887. Inspired by Enlightenment principles he maintained that "the universe is really a mechanism of the highest order." He had mathematical skills beyond those of other amateurs – indeed he was offered a life in professional astronomy. Tebbutt started to attract public attention when still young. By 1864 he had published an article on

John Tebbutt from the $100 note (1984)

Tebutt was best known for his discovery of "the Great Comet of 1861".

sunspots in *The Sydney Morning Herald*, and the popular press continued to maintain public interest in his work.

His maths skills were proven when he calculated the time of the solar eclipse in 1857. His first comet observation was the brilliant Donati's Comet of 1858, the orbit of which he calculated. International fame came when he discovered "the Great Comet of 1861" which he announced before it was reported elsewhere. He calculated its passage towards the Earth, predicting when it would appear in Europe. He continued to detect comets and calculate their orbit – until the return of Encke's Comet that had brought fame in 1822 to Carl Rümker. Between 1864 and 1907 he continued observations on eclipses of the moons of Jupiter, found the positions of minor planets and double and variable stars, and recorded the decreasing brightness of Eta Carinae. He published 350 papers, nearly half on comets, and participated in local and European scientific circles. Through the latter part of the 19th century John Tebbutt promoted the international credibility of observational astronomy. Memory of his work is perpetuated in the 21st century by a portrait on the $100 note.

Walter Duffield (1879–1929)
Richard Woolley (1906–86)
Clabon Allen (1904–87)

The excitement of astrophysics in the last decades of the 19th century followed the use of spectroscopy to begin analysis of a chemical universe. Asking basic questions of When? and How? Thus it bypassed a set of colonial observatories committed to largely utilitarian roles, while amateurs were locked into traditional observational astronomy. The efforts of one man led Australia into the new era. Walter Duffield, an Australian physicist, had trained in Cambridge and then Manchester in experimental spectroscopy with Arthur Schuster, a leading solar physicist. Previously in 1860, William Scott, Government Astronomer in Sydney, noted that "a connection between solar spots and magnetic variations, and the Aurora is due to or accompanied by magnetic disturbances." Scott's observation established an interest in, and value of, spectroscopic studies in an Australian location. This idea was of enormous importance and studies were also planned in northern observatories. But it was Duffield who saw the immediate opportunity for a solar observatory in Australia.

He began in 1905 a programme of promotion that included support from William Bragg (his academic link in Adelaide), lobbying through the Solar Union in Paris in 1907 and confronting opposition at home. A site was selected at Mt Stromlo, and in 1924 Duffield was selected as director of the Commonwealth Solar Observatory. His vision was to develop a portfolio of studies in modern astrophysics based on solar studies, the impact of the Sun on geophysics and spectroscopy of the Sun.

He died in 1929 before his vision could be realised, but that vision and the extraordinary amount of lobbying he did ushered in a new phase in Australian science, based on political process, large amounts of money and national interest. His contribution to research was not great, but he was able to move observational astronomy done by state astronomers and talented amateurs, into astrophysics and all that meant, by making it a national priority, sited within a new national capital, with a new look postgraduate university around the corner.

Duffield's death was followed by a period of research focussed on solar physics including correlating solar flares with sunspots and their causation of ionospheric disturbances known as "fadeouts" which would become of value in relation to radio transmission in World War II.

The most important work, and the studies that established an international reputation for Mt Stromlo, were Clabon Allen's photometric atlas of the solar spectrum. Allen continued his work to study dark lines, known as Fraunhofer lines, in the solar spectrum, due to specific absorption by elements in the Sun's atmosphere, which was important in stellar spectroscopy. In the war years Allen continued the work on "fadeouts" to correlate solar activity and Earth's magnetic field to forecast optimum radio transmission times. An extension of these studies previewed discovery of solar winds. Allen's contributions also led to recognition of interplanetary dust particles.

Richard Woolley was a British astronomer recruited to the directorship of Mt Stromlo Solar Laboratory in 1939, which at the time he recognised as a "modest establishment" but one with potential to develop into a major player in international astronomy. He recognised the value of its southern position which gave access to skies beyond the reach of observatories north of the Equator. His plan was to switch from solar phenomena, to stellar astronomy. After World War II he was faced with building essentially a new institution, and incorporated it into the Australian National University, where he was Professor of Astronomy from 1950. He wrote extensively promoting astrophysics, with his own interest in the dynamics of star clusters.

Following the War, the recognition that stars are powered by nuclear fusion, in which hydrogen atoms each gain a proton to become helium, gave new opportunities for research into star life cycles and the synthesis of heavier elements. In addition, Allen began cooperative work with the radiophysics laboratory, as the first collaboration between optical astomony and radio-astronomy. Allen was an outstanding solar astrophysicist born in Perth, who took the Chair of Astronomy in University College London, building it into a world class institution. His significant contributions in the early years of Mt Stromlo did much to establish its international status. He returned to England in 1957 to become Astronomer Royal having built the Mt Stromlo complex into an internationally competitive observatory of great value in scanning southern skies.

Joseph Pawsey (1908–62)
David Martyn (1906–70)
W (Chris) Christiansen (1913–2007)
Ruby Payne-Scott (1912–81)

Pawsey, Martyn, Payne-Scott and Christiansen represent a number of highly skilled physicists who were founding members of the CSIR group that evolved out of the radiophysics laboratory set up to study radar during World War II.

Joseph Pawsey provided strong and respected leadership of this group until his premature death from a brain tumour in 1962. Born and educated in Melbourne, he completed a PhD at the Cavendish Laboratory in Cambridge on an 1851 Exhibition Scholarship with S.A. Ratcliffe, a pioneer in radiophysics of the ionosphere. In Cambridge Pawsey studied the impact of the ionosphere on radio propagation, discovering irregularities in a layer of ionised gas 100 to 150 km above the Earth's surface which causes the reflection of radio waves upon which long range communication depends. He returned to Australia in 1940 to head the CSIR radiophysics laboratory, then involved in radar research. By the end of the War, and with a highly trained staff of 300, CSIR elected to keep the group together with Pawsey in charge.

He had noted secret reports of radar interference from radio waves possibly originating from the Sun. He developed antennae to record radio emission from the Sun that suggested solar temperatures of the order of a million degrees Celsius. These observations of the "quiet sun", as he designated it, were the beginning of radio astronomy. To obtain better discrimination, Pawsey recorded from cliff tops two signals – one direct, the other reflected from the sea. By observing the two, he detected an "interference pattern", which enabled the size and position of the source of radio waves to be identified. The strong signals correlated with sunspots. Interferometry (the term given to techniques where waves are superimposed causing interference, the analysis of which gives information) became the fundamental principal of many radio telescopes, including the Australian Telescope of Narabri in 1988, and its use by Pawsey to identify the size and source of radio emissions, was his most important contribution.

Ruby Payne-Scott with Alec Little and Chris Christiansen at Potts Hill

Working at Dover Heights and Potts Hill in Sydney between 1946 and 1951, Ruby Payne-Scott was central to studies that pioneered the new science of radio physics, with important observations on sunspot activity using interferometry. She developed with Joseph Pawsey formulae for Fourier Analysis of radio waves that led to the basic methodology of Fourier synthesis, strategies that continue today.

David Martyn was a Scottish physicist, who joined the CSIR's Radio Research Board and studied local fading of radio signals and the Luxembourg effect (where one radio signal in the ionosphere interferes with another). He noted that radio signals to the ionosphere induce changes of electron velocity that correlate with the absorption power of the ionosphere. This led him to an examination of upper atmosphere conditions, identifying temperatures of up to 1,000 degrees Celsius, and noting correlations of ionosphere disturbances with solar phenomena. In 1944 Martyn joined the staff at Mt Stromlo, bringing optical astronomy and radio physics together. With further observations of the upper atmosphere he was able to correlate changes in the ionosphere with variations in the Earth's magnetic field. He noted "solar tides" that create "winds" apparent in electron motion along lines of magnetic field.

Chris Christiansen was an innovative Australian physicist who thrived on taking on "big questions" and big projects, that would have important long term outcomes. He joined Pawsey's CSIR group in 1948, and immediately became interested in Pawsey's observations of radio emissions from the Sun and the enhancement of signals due to sunspots. Christiansen pioneered an array technique, observing solar eclipses from widely spaced positions to determine the site and size of solar radio emitting regions. He set up a number of antennae to produce a series of narrow beams of radio signal which enabled scientists to look at components of the Sun's emissions adjusting the antenna directions to follow the Sun. He mapped the distribution of radio emissions on a two dimensional contour map – the first plotted using Earth-rotation aperture synthesis (Earth rotation aperture synthesis is a form of interferometry that magnifies images by mixing signals from a collection of linear arrays that use the Earth's rotation to change the relative orientation of the arrays and the radio source). It shows the distribution of strong radio emission areas in the Sun's corona, and that the intensity of emissions is not circularly symmetrical. He used aperture synthesis based on two rows of antennas, again using the Earth's rotation to follow the Sun. At this stage Christiansen was leading the world in synthesis astronomy but he had insufficient computing facilities to handle the data analysis. A team at Cambridge was able to use the computer there to Fourier transform interferometer data – and won a Nobel Prize!

Christiansen's most important contributions to radioastronomy were, first, his innovations in instrumentation, beginning with the construction of the Grating Array (a unique array of radio telescopes arranged in a way that gave high resolution data – it was constructed to give both one and two dimensional maps of emissions from the solar surface between 1951 and 1957) and the Earth's Rotational Synthesis, at Potts Hill near Sydney. Next came his crossed grating multi-element interferometer at Fleurs, known as Criss-cross. It was later modified, to produce the Fleurs Synthesis Telescope. The Galactic and extra-Galactic astronomical observations made with these instruments were his second great contribution. In 1951, after Edward Percell of Harvard University detected the interstellar spectral line of neutral hydrogen, Christiansen rapidly confirmed Percell's observations and produced the first hydrogen line map, with the line extending around the Milky Way to show the galaxy's structure. Christiansen went on to take charge of electrical engineering at Sydney University, where he continued his innovative and observational studies and set about attracting postgraduate students to the study of big questions in radioastronomy.

Ruby Payne-Scott deserves special mention (and is further discussed on p.175). In 1941 she joined the CSIR Division of Radio-physics as part of the war effort, to research small signal visibility on radar displays and sort out difficulties with noise factors. The group, run by Joseph Pawsey, became interested in reports of extra-terrestrial radio signals, a phenomenon not considered possible at the time. With Pawsey in 1944 Payne-Scott conducted the southern hemisphere's first radioastronomy experiment with long wave radiation. After the War the group focussed on solar noise and sunspot activity, and emission from the solar corona.

Her particular interest was in the structure and character of non-thermal emission, and she was involved in studies identifying million degree temperatures in the corona, the association of enhanced radiation with sunspots and their location using interferometry, and measurement of delays in arrival times of bursts passing through a decreasingly dense coronal atmosphere.

Payne-Scott's most lasting contribution was her demonstration, with Pawsey, that the distribution of radio-brightness across the solar disc could be treated mathematically as a two-dimensional sum of an infinite series of simple waveforms known as Fourier Components, and therefore could be computed using a Fourier transform algorithm – this became a foundation principle of analyses in radio astronomy.

Physics

Physics, astronomy and mathematics are intricately associated components of science. To understand them, great minds of all societies from the times of Archimedes onwards have been obliged to string them together within the bounds of natural philosophy.

Proposals to separate the three entities and challenges to the blurring of borders between them are put to rest in appraisals of the remarkable Square Kilometre Array in Western Australia, arguably the most exciting challenge today in Australian science. It delivers an unimaginable data load for analysis by detecting low frequency radiowaves to identify red shifted radiation from neutral hydrogen, as a probe to understand the most fundamental events in the early universe. In this context, the bedrock of physics is the description of our world and the universe as a giant cosmos behaving in a singular, logical and predictable way, expressed in a set of mathematical formulae that we call laws.

Galileo began a period of observation, experimentation, and analysis using mathematics, culminating in Isaac Newton's laws of motion and universal gravitation, published half a century later, in 1675, in his *Principia*. These laws served science well for over two hundred years, accommodating great discoveries of terrestrial magnetism (William Gilbert, 1600), planetary motion (Johann Kepler, 1609), comet behaviour (Edmond Halley, 1705), conservation of matter (Lavoisier, 1781), wave theory (Thomas Young, 1801) and atomic theory of matter (John Dalton, 1803). However, with discovery of the electromagnetic wave spectrum (James Maxwell, 1865), quantum theory (Max Planck, 1900), x-irradiation (Wilhelm Röntgen, 1895) and radioactivity (Henri Becquerel, 1896), and early studies of atomic structure at the Cavendish Institute (J.J. Thomson and Ernest Rutherford), a new set of laws was required. They came first with Maxwell's laws explaining the behaviour of wave forms in the electromagnetic spectrum, and the quantum theory of energy packets by Max Planck (1900). Early in the 20th century as the nascent ideas on subatomic physics developed, Albert Einstein added

energy into these equations – first, in his special theory of relativity in 1905 with his postulate of mass-energy equivalence ($E = mc^2$), and ten years later by adding gravity, in his general theory of relativity.

Such was the overwhelming importance of Newton's laws to the discipline of physics and its new definition as the "study of matter and energy", that all disciplines of science sought an underpinning cohesive principal – a single encompassing "law" – to legitimise their identity. For example, the classical life sciences developed classification, then Darwinian evolution, and finally the genetic code identified by Crick and Watson in 1953. Geology split from biology within the natural sciences aided by acceptance of the Plutonic theory of geological cycles, then discovery of plate tectonics as a defining law. Chemistry came of age with the progressive recognition of a regular periodic table relating the behaviour of chemicals to their structure.

The universal and fundamental impact of this developing understanding of nature promoted a broader community interest in science, particularly as its relevance to every day events was appreciated. It was a major component of the Enlightenment which, as discussed elsewhere, determined the course taken by science in Australia. Immigrants arrived with a sense of enquiry and belief in their capacity to solve problems, and the new society adopted the ideas and questions of the day, and expressed them in a remote and different land with its own demands and time frame in which problems had to be solved scientifically.

The discipline of physics was central to resolving challenges posed by navigation – especially astronomy, meteorology and magnetic deviation, which have been discussed. Physics was also part and parcel of the problems facing the new Australians, in agriculture and survey. These applications have been discussed already. Here we examine the traditional discipline of physics as the study of matter.

The development of classical physics in Australia is best traced through academic departments. In colonial times and the early years of Federation, four universities had departments of physics – Sydney, Melbourne, Adelaide and Tasmania. Physics developed later in Brisbane (where the first professor was Thomas Purnell, 1919–48) and Perth (with foundation professor of physics A.D. Ross, 1913–51). In both those places large teaching loads and limited resources, delayed significant research until after World War II.

In Adelaide, Sydney, Melbourne and Tasmania physics began as a foundation discipline – part of "natural philosophy" and associated with mathematics. The early professors were English, young, with outstanding academic records, usually obtained in Cambridge. Conditions in colonial universities were primitive and teaching loads huge. Those recruited to Adelaide were most successful in establishing a research culture. Horace Lamb (1849–1934) and William Bragg (1862–1942) were outstanding in their times. In all the universities were long periods, usually within the tenure of a single professor, during which teaching and administration loads grew heavy and resources and funds were lacking, so that little research of international standing was completed, although many research programmes of moderate importance were begun. The World War II years involved the university physics departments in activities geared to the war effort. The immediate post-war period was a watershed, with a new wave of people and ideas, and more resources.

Of particular importance was securing Sir Mark Oliphant to develop physics in the new Australian National University. His great cyclo-synchrotron accelerator project that he anticipated would take Australian nuclear physics to the forefront of world research was never completed. But he recruited outstanding scientists and broadened the programme to include astronomy, mathematics, geophysics, theoretical physics, atomic and molecular physics, nuclear physics and particle physics. The undergraduate universities followed and listed similar topics as research interests.

Earlier in the 20th century bright young Australian physics graduates, with little local research to attract their interest and no PhD programmes available, took advantage of scholarships, in particular the 1851 Exhibition Scholarship and Rhodes Scholarships, to get postgraduate training in Britain. As in other disciplines, young Australians aspiring to work in physics were attracted to research groups run by English academics who had worked in Australian universities, such as William Bragg, or by physicists from Australia and New Zealand with established reputations, such as Ernest Rutherford at the Cavendish Institute in Cambridge.

Later Mark Oliphant had John Gooden and W.I.B. Smith from Adelaide, John Blamey from Melbourne and Len Hibbard from Sydney working in his Birmingham laboratory. While it is true that the greatest achievements in theoretical physics were by those who returned to or stayed in English institutes such as the Cavendish, University College, London (where William Bragg worked), Victoria University, Manchester (William Lawrence Bragg) and Birmingham (Mark Oliphant), like Australian scientists in other disciplines, Australian graduates were well accepted, and contacts with Australia well maintained. Thus, in the immediate post-War period, Australian physicists returned to reinvigorate university departments, to be part of the Australian National University where research was the main focus and to join organisations including CSIR which was establishing research in astrophysics and other areas.

The story of physics in Adelaide up to the mid-20th century, is the story of Horace Lamb, William Bragg and Kerr Grant. Lamb and Bragg were young Cambridge graduates in mathematics whose interest in physics grew during their time in Adelaide. Both made an effort to foster research during their stays, of nine and twenty two years respectively. Lamb published twenty research papers from Adelaide. In 1883 his application of Maxwell's equations to oscillatory current flow in spherical conductors was the first published research in Australia in theoretical physics.

Lamb was followed by the twenty three year old mathematician William Bragg, who made an impact on Adelaide by his efforts to promote science and his

Oliphant's Homopolar Generator, Australian National University (1953)

Oliphant's plan for a world leading nuclear physics laboratory in Canberra was based on constructing a synchrocyclotron to make and study anti-matter. This homopolar generator was to be the power source to drive the synchrocyclotron, which in turn accelerated charged particles such as protons and deuterons.

involvement in local and national scientific societies. His physics research related to electromagnetic waves and radiation including radiotransmission, imaging with x-rays, and absorption of alpha, beta and gamma rays. Later he studied ionising radiation and its interaction with matter. According to his son, with whom he would share the Nobel Prize for their development of x-ray diffraction to identify crystal structure, the work in Adelaide initiated the ideas that led to his work on crystals.

The third professor of physics was Kerr Grant, Australian born, and educated in Melbourne. He held the position from 1911 to 1948. He saw his role as primarily one of teaching, and little quality research came from Adelaide during this period. Two exceptions should be noted. First, Roy Burden, an academic in Kerr Grant's department for much of his tenure, was a good scientist with limited opportunities for research. His work on the surface tension of mercury was important, and he supported and influenced many graduates, a quarter of whom would go on to research careers.

The outstanding graduate in this time was Mark Oliphant, who wrote in a positive way regarding the academic value of Burden. Oliphant would become a pioneer in nuclear physics working on an 1851 Exhibition Scholarship with Ernest Rutherford at the Cavendish Institute in Cambridge in the 1930s. His classic studies on nuclear fusion and recognition of the potential for chain reactions, were critical to the construction of the atomic bomb in the U.S.A. When Rutherford retired, Oliphant took a chair of physics in Birmingham in 1937, with the aim of building a large cyclotron. From there he was recruited into the radar development programme as part of the war effort. After 23 years, Oliphant returned to Australia to participate in the development in 1950 of a School of Physics at the Australian National University that set foundations and became a model for academic physics in late 20th century Australia.

In Adelaide by the mid-1940s, the great tradition established by Lamb and Bragg had been lost and there was little research activity. That would change in 1948 with the appointment of Leonard Huxley, a Tasmanian who had been to Oxford on a Rhodes

Scholarship, then worked with Oliphant in Birmingham on electron motion in gases and radiowave propagation in the atmosphere. In Adelaide Huxley began research programmes in electron diffusion in gases, radiowave propagation in the atmosphere, and radar meteor astronomy. The department would participate in upper atmosphere studies in Antarctica as part of the International Geophysical Year in 1958. To renovate the department, Huxley recruited Stan Tomlin and Harry Medlin from England. A second chair in mathematics and physics created in 1949 was filled by Bert Green, who was largely responsible for the development of a plethora of interests in "modern physics" that would characterise departments across the country in the latter part of the century.

Melbourne and Sydney Universities taught physics from their inception, as "natural or experimental philosophy", but with little research. Teaching and administration loads were high and teaching usually involved additional disciplines. In 1915 Professor Thomas Laby was appointed in Melbourne, where he developed a dynamic department still with little research. He was followed by Professor Leslie Martin (from 1945 to 1959) who had studied at the Cavendish with Rutherford in nuclear physics and brought new ideas to the department including setting up the first electronic computer in an Australian university, CSIRAC, in 1955.

In Sydney, Richard Threlfall was appointed as the first professor of physics in 1886. He had worked at the Cavendish with J.J. Thomson and established the first quality laboratory facility, known for accuracy of measurement, and an active research programme. He was followed by James Pollock (1899–1922) and then Oscar Vonwiller, in turn followed by Harry Messel in 1952. At the time of Messel's appointment, little significant research was being done. Vonwiller studied optics, Briggs who had studied with Rutherford had a research programme in nuclear physics and Bailey's studies on the ionosphere led to the discovery of the Luxemburg Effect (interference of short wave wireless signals). Messel, a Canadian, had come from Adelaide where he worked on cosmic rays. He rejuvenated a tired

department establishing a School of Physics, industry-based funding and SILLIAC – an electric computer. Sydney and Melbourne physics departments went on to become "modern" schools along the lines of Oliphant's school in Canberra, covering the widest range of interests.

Physics in the University of Tasmania was dominated by the father and son McAulays. Alexander McAulay (1863–1931) was an English trained mathematician with an interest in physics, appointed as the foundation professor of physics in 1896. A practical man, focussed on relevant problems, he completed the first magnetic survey of Tasmania, and a valuable text on log tables. He began practical experiments on hydrodynamics which led to the first major hydroelectric scheme in Tasmania at Waddamana in 1916. He was followed in the chair by his son, Alexander Leicester McAulay (1895–1969) who had a more traditional training with Rutherford in the Cavendish. After his father's retirement, he became professor of physics in 1927, bringing a very modern approach at least in his range of interests – particle physics, cosmic rays, and metal surface electrochemistry. He became particularly interested in biophysics, with studies in Tasmania and Antarctica.

The McAulay practical bent was evident in World War II, when he joined many academics in the developments of precision lenses and gun sights using innovative methodology.

Not all Australians studying nuclear physics in pre-war Britain returned to academic positions in Australia. Peter Thonemann (1917–2018), born in Melbourne and with a BSc (Melbourne) and MSc (Sydney), moved to Oxford where his post-graduate work in toroidal discharge (involving circular currents around a central space) led to critical studies in nuclear physics involving the design and construction of a toroidal device ZETA to heat and magnetically confine ionised deuterium to control the release of fusion energy (with parallel studies in Britain developing a hydrogen bomb, based on uncontrolled release of fusion energy). Thonemann later took a chair of physics in Wales, switching his research to biological outcomes of nutrient gradients in *E. coli* bacteria.

Important Physicists

Lawrence Hargrave (1850–1915)

A complex and solitary man, largely self taught in mechanics and aerodynamics, Lawrence Hargrave lacked only an engine with an appropriate power-weight ratio, to be first to fly a heavier than air aeroplane. Indeed, aeronautical engineering students in 1992 used Hargraves blueprints to construct a plane, which fitted with a modern engine, flew!

A gentleman inventor, well off through inheritance and with a love of adventure, nature and wonder, Hargrave's aim in life was to invent, build and fly an aeroplane. Initially trying to copy nature with flapping wings, by 1890 he had identified key components for successful flight. First, he showed that a cellular box kite wing had superior stability and soaring power due to an improved lift to drag ratio, that a curved wing surface

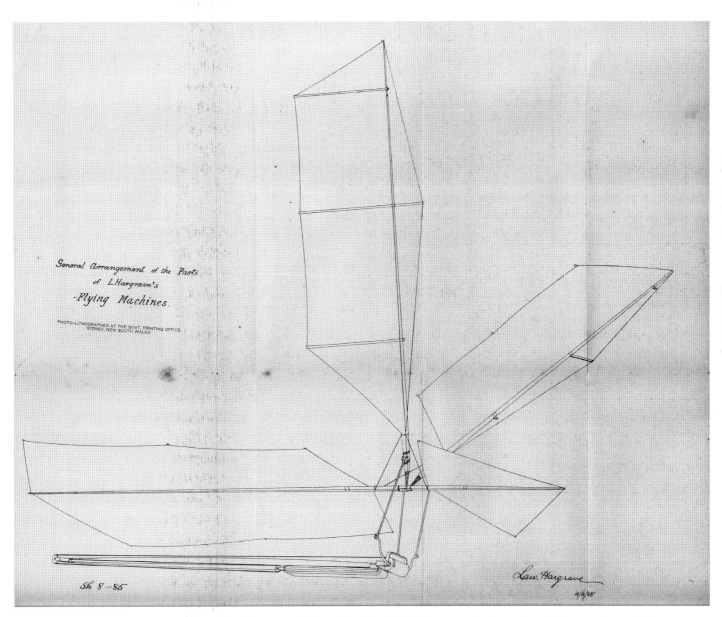

"General Arrangements of the Parts of L. Hargraves Flying Machines", L. Hargraves in *Journal and Proceedings of the Royal Society New South Wales* (1885).

Hargraves published most of his scientific studies on details of his "heavier than air flying machines" in the *Journal of the Royal Society of New South Wales*, without patenting his inventions.

If there be one man who deserves to succeed in flying through the air, that man is Lawrence Hargrave of Sydney.

Octave Chanute, *Progress in Flying Machines*, 1893

gave twice the lift of a flat surface, and that a thick leading wing surface (or aerofoil) was optimal. Second, he developed a revolutionary radial, rotating engine driven by compressed air, which was later adapted and used in commercial and military planes for several decades. Voisin constructed the first commercial aeroplane in France, which he called a Hargrave to record his respect for the Australian's achievements. Hargrave believed in free information exchange and never patented his inventions – he published all his work, mainly in the *Journal of the Royal Society of New South Wales.*

"I know that success is dead sure to come," he wrote. It did, but to the Wright brothers in 1903, who almost certainly copied Hargrave designs which had been published in Octave Chanute's *Progress in Flying Machines.*

Horace Lamb (1849–1934)
William Bragg (1862–1942)
Lawrence Bragg (1890–1971)

Three remarkable mathematicians/physicists kickstarted academic physics in Australia. Horace Lamb and William Bragg were the first and second Sir Thomas Elder Professors of Mathematics in Adelaide. Both graduated from Cambridge in mathematics, with no clear future in British academia, and no clear direction as to how they would develop their careers. William Lawrence Bragg, William Bragg's son, born and trained in Adelaide, completed his first degree in his father's department.

Lamb arrived in 1876, a month before the inauguration

of Adelaide University, as one of the founding professors. He established a dynamic department, and wrote *A Treaty on the Mathematical Theory of the Motions of Fluids.* Published in 1879, this popular work was reprinted as *Hydrodynamics* up until World War II. He began a series of analytical applied mathematical studies on topics such as flow in conductors, sound wave transmission and infinitesimal calculus. He left Adelaide in 1885 to take the chair of mathematics in Manchester.

Coming from humble beginnings, William Bragg took the Elder Chair to create for himself a future unlikely to be available in England. With little knowledge of physics, he found that he was to teach that discipline, and that he enjoyed it. He adjusted easily to expanding social and professional experience. He saw science as a community issue and became involved in promoting it at many levels. He became interested in electromagnetic waves and radioactive particles. At the 1904 meeting of the Australian Association for the Advancement of Science he gave an important paper on "Ionization of Gases", describing how the behaviour of alpha and beta particles differs from x-rays and y-rays, with alpha particles losing energy in ionising gases. Prior to leaving for Leeds in 1909 he summarised his studies, indicating that the stopping power of particles is proportional to the square root of the atomic weight of the substance, and controversially stating that x-rays and y-rays are streams of neutral pairs of particles, rather than electromagnetic waves.

During William's 23 years in Adelaide, his son William Lawrence grew up and graduated in Bragg's department. In 1909 they left for England. After Laue, in 1912, showed that crystals diffract x-rays and confirmed the wave nature of x-rays, William developed a spectrometer to measure the diffraction more accurately, while Lawrence determined Bragg's Law, which correlates the maximum diffraction pattern with the wave length of the x-rays and the distance between the planes of atoms in the crystal. Lawrence appreciated that the diffraction pattern allows recognition of the site of atoms within the crystal. Experimentally these ideas were tested on crystals of zinc sulphide and alkaline halides. With improved methodology devised by William, the era of x-ray crystallography was born.

During World War I William worked on submarine detection projects while Lawrence studied sound ranging for artillery. Both were later awarded Nobel Prizes, knighted and elected to the Royal Society.

Lawrence Bragg succeeded Rutherford as head of the Cavendish Institute in Cambridge. Lawrence expanded his interest in crystal analysis with inorganic substances and after his move to the Cavendish, focussed on the study of protein structure, using x-ray crystallography while taking advantage of the computing facilities at the Royal institution, to identify the structure of myoglobin, haemoglobin and lysozyme. They maintained connections with Australian science, returning to Australia on visits and training postgraduate students.

Mark Oliphant (1901–2000)

The most influential Australian physicist was born in Adelaide and completed his BSc in physics in the department begun by Horace Lamb and William Bragg, inspired by the flamboyant Kerr Grant but impressed by Roy Burden the respected scientist. Work on surface tension of mercury with Burden led to Oliphant receiving an 1851 Exhibition Scholarship to work with J.J. Thomson at the Cavendish Institute in Cambridge. Following completion of his PhD in neutralisation of positive ions on a metal surface, he began the most exciting and productive period of his career. In 1932 eight scientists at the Cavendish had Nobel Prizes or would later get one. It was the cusp of the age of nuclear physics based on Rutherford's model of the nucleus, complemented by the discovery of neutrons by

Chadwick. Rutherford recognised Oliphant's particular skills in constructing laboratory equipment.

Two critical studies were completed: separation of potassium isotopes by centrifugation, and transmutation of elements bombarded with protons, following Cockcroft's demonstration that when lithium is bombarded this way, showers of protons and helium nuclei are released. Cockroft had built a high voltage particle accelerator. The studies revealed that bombarding heavy water with heavy hydrogen (or deutons) leads to fusion then disintegration (the basis of nuclear fusion energy).

The future of nuclear physics seemed unlimited. But by the late thirties Chadwick and Rutherford had left, Lawrence Bragg was director shifting the focus of the Cavendish towards solid state physics, and Berkeley in the U.S.A. had taken the lead in nuclear physics with the most powerful cyclotron. Oliphant accepted a position at Birmingham, as World War II loomed. He was asked to join the war effort by improving radar at the time when the model was crude and bulky. His recognition of the need for shorter wave length radiowaves made portable units possible and his production of a resonant cavity magnetron revolutionised the Allies' use of radar by miniaturising equipment. Every ship and plane could then carry a radar unit.

He was aware that a German group had bombarded uranium with neutrons and found, contrary to experience with light elements, that transmutation produces smaller atoms due to fusion, with release of energy. In France, it was noted additional neutrons remained, making possible a chain reaction. Oliphant and his group began enrichment of uranium to obtain the isotope U235 in pure form to make an atomic bomb a realistic outcome, by using the centrifuge system he had previously developed to isolate potassium isotopes. The scale of the operation was greater than then possible in Britain, so Oliphant moved his team to America, to join the group at Berkeley.

After the War, Mark Oliphant became committed to the peaceful use of nuclear energy, and after twenty three years in England accepted an offer to build a high energy accelerator physics unit within the Australian National University. He would develop and run one of the four post-graduate research schools, setting a model followed by state universities by adding wide ranging project areas to mainstream physics, including astronomy, geophysics and mathematics.

His ambition for the school to build an accelerator operating at 2GeV (twice the energy of the Birmingham proton synchrotron) was thwarted by international competition. Oliphant was a powerful influence on the development of science in Australia after 1950, playing a leading part in developing a scientific community through the Australian Academy of Science, and acting in numerous national and international advisory roles.

Thomas Laby (1880–1946)
Les Martin (1900–1983)
Len Huxley (1902–1983)

Academic physics in Australia was in the doldrums in the late 1940s. For decades tenured heads of department had struggled with poor resources and high teaching loads, with little progress to show. The exception was Thomas Laby. Born and trained in Australia, he won an 1851 Exhibition Scholarship to work on ionisation by alpha particles with J.J. Thomson at the Cavendish Institute. After positions in England, Sydney and New Zealand, he took the chair of Natural Philosophy in Melbourne in 1915. He took on every role and every position in professional societies, university committees and the community, promoting and organising science and physics with commitment and enthusiasm. He continued his research in studies of radioactivity, especially its medical applications. However his research was not well focussed and spread into wave mechanics, geophysics and optical studies. Always promoting research, he sent twelve postgraduate students to Cambridge on 1851 Exhibition Scholarships. It is not surprising that following retirement in 1942 he was succeeded by one of his students who had studied with Ernest Rutherford in the Cavendish, supported by an 1851 Exhibition Scholarship.

Les Martin who followed Laby as Professor of Physics in Melbourne, had held junior positions in the department from 1927. His early work in x-ray characterisation switched to nuclear physics, targeting accelerated deuterons onto heavy water to generate fast neutrons. During World War II he was seconded to the CSIR radiophysics laboratory. Forced to build his own betatron, electron synchrotron and various accelerators and cyclotrons, Martin was never able to create an internationally competitive research unit, partly because of his lack of experience and a massive administration load within the university and national bodies concerned with atomic physics. This concluded with an active period for him, when Australia was hosting British atomic weapon testing and he was able to share in the intense interest in the wider aspects of radioactivity. He went on to be a leader in driving Australia's tertiary education sector. His success was based on his integrity, trust and political skills.

Physics in Adelaide could hardly have had a better start than it got with Horace Lamb and William Bragg. But there were few research initiatives under the long tenure of Professor Kerr Grant. That would change with the appointment of Len Huxley to the Elder Chair of Physics in 1948. Huxley had grown up in Tasmania and completed a degree in physics at the University of Tasmania. He was awarded a Rhodes Scholarship to Oxford, where he completed a PhD with John Townsend on electrical breakdown in gases. Electron movement in gases became a long research interest for him. He moved back to Australia, joining the Radio Research Board before returning to England in 1932 to work with Oliphant. After radar research in the war years, he moved to Birmingham, and from there was recruited to Adelaide.

When he arrived, the new Vice Chancellor, A.P. Rowe, was invigorating a university in decline and federal money was becoming available. Huxley recruited staff and created PhD programs and new research directions. He established a radar group tracking meteors and upper atmosphere winds, while continuing his work in electromagnetic wave dynamics in the ionosphere through another team. Others in the department developed programmes in geophysics, solid state physics and seismology. The time in Adelaide was Huxley's most productive research period, with studies on the Luxembourg Effect (which occurs where one radio signal interferes with another) and laboratory studies of electron motion in gases. He described trails of ionised gases left by meteors in relation to wind patterns in the upper atmosphere.

His subsequent appointment as Vice Chancellor of the Australian National University ushered in a new phase where his feel for what a university is about blended with a high level of administrative skill, enabling a continued significant input into science and academia at a national level.

Transformation of physics departments in other states would come later. Sydney University physics under Professor O. Vonwiller (1923–46) had fallen away, but the recruitment of Harry Messel would change that. Messel had studied under Erwin Schrödinger who developed the quantum model of the atom. He had an immediate impact, recruiting new physicists, developing novel funding schemes and advanced computing while pursuing his own interests in tracking crocodiles with remote sensors in Queensland estuaries.

In Tasmania, following the retirement of Alexander McAulay in 1960, Professor Graham Ellis developed astrophysics as the core area of study, with his own expertise in monitoring low frequency radio waves, determining their source and relationship to magnetic fields and auroras. Prior to 1950 little research came from physics departments in Western Australia and Queensland.

Engineering

Classical, basic scientific studies in physics and mathematics stood behind the scientific achievements of pre-1950 Australia in areas of applied science related to national growth — including engineering feats building the infrastructure of a vast, poorly populated land.

All nations have engineering projects but Australia was set apart by the immensity of what was achieved in a short time frame, very often with unique challenges in remote and hostile environments.

Applied physics, or engineering, involved innovation in response to need, "the mother of invention", often driven by the notion "there's got to be a better way". Extraordinary feats in engineering were seen as part of the fabric of science; many articles by engineers were published in the journals of royal societies, and the Australasian Association for the Advancement of Science had a separate division for engineering.

Many examples appear elsewhere in this book, and others are covered here. Contributions by Australian engineers using scientific methodology ranged from massive infrastructure projects such as the largest steel through arch bridge in the world, across Sydney Harbour (opened in 1932) and the Snowy Mountains hydroelectricity and irrigation scheme, one of the most complex integrated systems in the world (begun in 1949), to mechanical inventions that contributed to industrial processes in primary and secondary industries. Nowhere is the idea of "science owned by the public" more evident than in what broadly passes as "engineering" where small businesses showed invention and innovation and use of scientific method – most not remembered or recognised in today's world.

For example, William Clancy built a repetition engineering business, designing and making machinery based on rotating metal rods operated on by tools controlled by a camshaft. In the late 1940s Clancy designed and built a multi-axis machine enabling three dimensional processes to operate on raw bar stock, the bar remaining stationary while the camshaft worked rotating tools that superseded manual secondary processing. It was a major cost cutting advance! Clancy

Charles Kingsford Smith's famous *Southern Cross*, followed by *Clancy Skybaby* (1933)

The "Clancy Skybaby", designed and built by William Clancy, was the first registered aeroplane designed and built in Australia. The design included innovative features later incorporated into commercial planes especially in engine design. He extensively modified a motorbike engine to achieve adequate power to weight ratios, replacing the top half of the engine with new cast cylinders and alloy pistons, and converting a side valve system to an overhead valve engine.

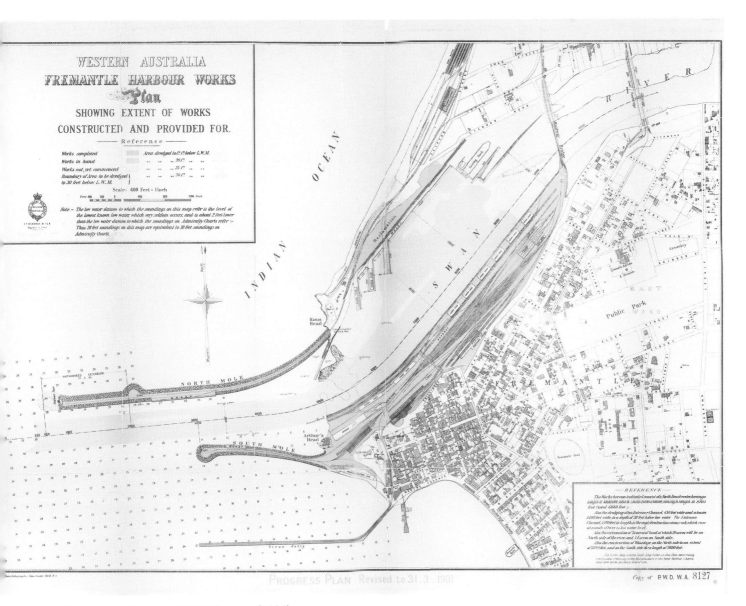

Fremantle Harbour, C.Y. O'Connor (1892)

Defying contemporary wisdom O'Connor created a durable and safe deep water harbour at the entrance to the Swan River. This involved removing a limestone bar and excavating sand shoals.

is better known for designing and constructing the first registered home built aeroplane, the Clancy Sky Baby, now in Sydney's Powerhouse Museum. The plane's innovative mechanics included use of overhead valves in a modified engine and years before similarly modified engines were used in commercial planes. To obtain the one thing lacking in Lawrence Hargrave's designs – an engine with appropriate power:weight ratio, he expanded the cylinder size of a Henderson motorbike engine.

Science based engineering built today's Australia

by construction (of tunnels, railways, roads, towers, bridges and buildings), retention and control of water usage (including the Snowy Scheme, O'Connor's goldfields water supply scheme to Kalgoorlie, schemes for irrigation and sewerage reticulation schemes and dams), communication such as the Overland Telegraph, 3000km of single iron cable reducing communication times to the opposite side of the Earth from months to seven hours in 1872).

John Whitton supervised the construction of over 2,000 miles of railway in New South Wales, while

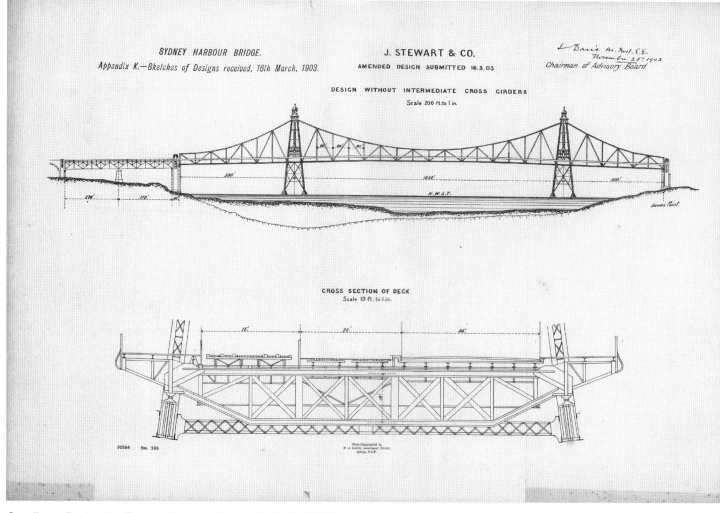

SYDNEY HARBOUR BRIDGE.
Appendix K.—Sketches of Designs received, 16th March, 1903.

J. STEWART & CO.
AMENDED DESIGN SUBMITTED 16.3.03

Chairman of Advisory Board

DESIGN WITHOUT INTERMEDIATE CROSS GIRDERS
Scale 200 ft.to 1 in.

CROSS SECTION OF DECK
Scale 10 ft. to 1 in.

Cantilever Design for Sydney Harbour Bridge, N. Selfe (1902)

Norman Selfe was the dominant engineer in late colonial Sydney, involved in every aspect of his profession including teaching. His cantilever design won the 1902 bridge competition, and would have become the harbour crossing but for a shortfall in funding. This was his winning design. The later switch to an arch bridge only became possible due to invention of altered metal properties.

Lawrence Hargrave's pioneering studies made possible heavier than air flight. In the mining industry there was ore refining by flotation and cyanide extraction for gold. In mechanical engineering, inventions ranged from the stump jump plough and the combine harvester, to Arthur Bishop's variable ratio nose wheel steering system followed by variable ratio rack and pinion steering, first used in Australian cars.

Many "great Australian engineers" were born overseas – in Great Britain, Ireland and France. Some came as part of a European diaspora in the second half of the 19th century, while others were recruited for major infrastructure projects. Many were involved in scientific societies and some were elected as fellows of the Royal Society (London) including William Hudson and Anthony Michell (for his discovery of thrust technology enabling the building of large ships). Some who made their mark were trained in Australia, including Michell and Ronald Bracewell (an astrophysicist with inventions related to wave mechanics).

Growth in scientific engineering and the need for it reached crisis point in the 1880s in Australia. Melbourne had begun university education courses in engineering in 1861, stimulated by wealth and development that followed the goldrush of the early to mid 1850s. In Sydney, the Sydney Mechanics School of

Arts had begun to offer courses providing a scientific foundation to technical training in 1878. Within three years there were more than a thousand students in over 50 courses, which included basic engineering subjects such as mechanical drawing, applied mathematics, and steam engine science. This extraordinary response got the government and the universities' attention. In 1883 the government took over the SMSA programme, moving it in 1891 to a permanent site in Ultimo to become the Sydney Technical College. Meanwhile the University began teaching engineering subjects. Between 1891 and 1959 the College was the centre of science-based technology in New South Wales, and gave diplomas in engineering. Across the colonies technical education developed with a particular focus on mining and engineering – these included the Ballarat School of

Mines in 1870 (the third oldest academic institution in Australia) and the South Australian School of Mines and Industry in 1889.

Two generalist engineers in Sydney at that time were P.N. Russell and Norman Selfe. Russell built up the largest engineering company in New South Wales and was instrumental in starting the engineering school in Sydney University by donating funds. Selfe was an extraordinary innovative and talented engineer, competent in a wide range of areas from shipbuilding to ice making machines, and with visions of projects such as city railways, a harbour bridge and an independent system of technical education – all of which came about. He was actively involved in the Royal Society of New South Wales and the SMSA, and was a founder of the Engineering Association of New South Wales in 1870.

Important Engineers

John Whitton (1819–98)
Charles O'Connor (1843–1902)
William Hudson (1896–1978)

To achieve the growth required to support nationhood, major infrastructure had to be constructed: railway networks for produce and human transportation, harbour construction to support international trade, and optimal management of water resources in a dry environment. Such tasks were beyond locally trained engineers with limited experience, so engineers were recruited to Australia. Whitton, O'Connor and Hudson became legends in contemporary international engineering, contributing far more than the tasks for which they were recruited. Each required great political agility and associations yet had also to understand and manage a large workforce and respect budgets when so much of the work involved unseen hazards. All would make Australia home and spend the rest of their lives in their adopted country. These were scientists of engineering.

John Whitton arrived in New South Wales in 1856, from England where he had supervised construction of railways. Appointed as chief engineer and charged with the construction of railways in New South Wales, he arrived to find 23 miles of railway, few steam engines

and no clear plan on how to proceed. When he retired 34 years later, there were 2,174 miles of track on which no accident had occurred due to defective design or construction. His innovative and creative construction of two zigzag rail inserts to cross the Blue Mountains was followed over a five year period in the 1880s with a thousand miles of new track that carried nine million passengers annually. He completed railways from Sydney to Bathurst, Albury and the Queensland border. Other challenges he took on included water and sewerage projects and plans for a harbour crossing in Sydney. He established workshops and railway infrastructure. He was energetically supported politically by the Premier, Henry Parkes.

C.Y. O'Connor was employed and supported by the Premier of Western Australia, John Forrest, in 1891 as the colony's engineer-in-chief to be responsible for "everything". They worked together for a decade to revolutionise infrastructure in the west. Forrest needed a port of international standard close to Perth, for mail and export of primary produce. There was doubt as to the possibility of building a reliable sheltered harbour due to a limestone bar at the entrance of the Swan River, and concerns that littoral sand travel would silt approaches. O'Connor was unimpressed with the studies available and did his own, concluding that there was no serious sand movement, that the bar could be removed, and the Harbour deepened and kept clear by dredging – all within Forrest's budget and timeframe. He compiled plans for an inner harbour for vessels drawing 9 m at low tide, for completion in eight years at a cost of £800,000. By 1900 the harbour was complete and receiving P&O mail liners from London. This was O'Connor's greatest achievement, though he is better remembered for preparing and prosecuting plans to supply water via a pump assisted pipeline of 330 miles with an elevation of 1,000 feet over an escarpment, delivering five million gallons daily to the goldfields from a reservoir in Coolgardie. Nothing of this magnitude had been constructed in Australia before. It was a critical part of O'Connor's and Forrest's plan to modernise the West, providing for the men and mines, the railways, and the townships.

O'Connor also restructured and organised Western Australia's railways including repair and maintenance workshops. The system began to turn a profit. The two edged sword of political support led to extraordinary attacks through the press on O'Connor once Forrest moved to federal politics, ending in his suicide in 1902.

William Hudson shared with Whitton and O'Connor experience in major infrastructure works related to water control and railways, an extraordinary work ethic and commitment to scientific engineering based on data collection and problem solving. He had spent two periods in Australia, in charge of the construction of the Nepean and Woronora dams, and developed a particular interest in hydroelectric schemes when supervising the construction of the Galloway scheme in Scotland. His appointment in 1949 as manager of the Snowy Mountains Scheme – the biggest engineering project ever in Australia – came after the federal Minister for Works, Nelson Lemmon, reportedly submitted three names for the position to Prime Minister J.B. Chifley, "Hudson, Hudson, Hudson". In achieving the redirection of east flowing water in the Snowy and Eucumbene Rivers through tunnels to supplement the Murray and Murrumbidgee in the west, the project produced water for irrigation while generating electricity for much of the Australian Capital Territory, New South Wales and Victoria. He managed a workforce of nearly eight thousand to build 16 dams, seven power stations, 50 miles of aqueducts and 90 miles of tunnels, completed early (in 1974) at a cost close to budget of £422 million. The Scheme generated 3.74 million kW of power annually and distributed 2.4 million megalitres of water for irrigation. Political support came through Prime Minister Robert Menzies. Hudson was a fellow of the Royal Society and awarded several honorary university degrees.

John Bradfield (1867–1943)

John Bradfield was one of the first Australian born and educated engineers to control a major infrastructure engineering programme in Australia, as supervising engineer for the construction of the Sydney Harbour Bridge – a project that had been promoted since the 1880s. He was more than that, with an outstanding academic record and long-term ties to the professional development of engineering. His early work contributed to the construction of the Cataract and Burrinjuck dams near Sydney and drove a broad set of ideas related to the development of Sydney infrastructure. He supervised electrification of suburban trains and a city rail circuit. He always promoted his engineering plans in the context of the broader community.

His great purpose from his student days was to build a bridge across the Harbour. He favoured a cantilever bridge (much along the lines of an earlier prize winning model designed by Norman Selfe), but after new developments in light steel made an arch bridge possible he made plans for both possibilities. He combined his ideas for a bridge and electrified railway in the first doctorate in engineering at Sydney University (1924). Fellow engineer John Monash described Bradfield's proposal as being "of exceptional magnitude – and unique in engineering practice". The resulting Harbour Bridge included inputs from a number of engineers, but the vision, the concept and its broad design, with its integration into the fabric of Sydney's transport infrastructure, were all Bradfield.

"Zig Zag Railway, Looking Westways" in *The Railway Guide of New South Wales* (1881)

John Whitton was the master engineer of the New South Wales railways, built at an astounding rate between 1856 and 1890, connecting Sydney with country areas. His masterpiece was the Great Zig Zag at the western end of the crossing of the Great Divide, built between 1866 and 1869. His development of locomotive boilers better capable of powering engines up steep slopes, became a model copied by mountain rail services throughout the world.

Anthony Michell (1870–1959)

Michell was born in Victoria, to English parents who had migrated to Australia in search of gold. After studies in Melbourne and Cambridge he settled in Melbourne and completed a degree in civil and mining engineering. He became interested in the physics of fluids, including viscosity and lubrication.

He discovered the advantage of separating moving metal parts with an oil film, which improved efficiency by reducing friction, allowing a tenfold pressure increase.

He applied this to thrust bearings in marine engines, replacing existing plane faced metal-on-metal with a tilting slipper pad; with oil introduced between the pads and a collar on the shaft, the load could be taken by the oil film. His innovation revolutionised marine propulsion, enabling the building of massive ships, the size of the *Queen Mary*. In what can be thought of as "par for the course", the idea was not understood or taken up by Australian or British interests until Michell's thrust bearing was found in a captured German submarine.

Michell adapted his ideas on oil film mechanics to develop a "crankless engine", aimed at reducing size and weight and improving fuel efficiency of the internal combustion engine. He reduced size by arranging the cylinders together rather than in line, with their pistons pushing in turn on a slanted swashplate which was pushed and pulled in a constant perfect wobbling pattern. The swashplate translated its motion to the central shaft to which it was attached. The pistons operated parallel and adjacent to the output power shaft, with an oil film separating the shoes and the disc. Despite being successfully demonstrated on Michell's own car, the system was never taken up by manufacturers.

Michell worked with Queensland engineer Louis Sherman to build a range of engines and pumps. His commercial success was limited but the quality of his scientific inventions was widely recognised and he was awarded a fellowship of the Royal Society of London.

Epilogue 2

Underpinning the foregoing section is the idea that the term "Navigational Science" gave an Australian identity and context for branches of science in the classic division known as "Natural Philosophy" which were encompassed within the package of scientific endeavour associated with James Cook. Those disciplines, survey, astronomy and physics, were the focus of the Enlightenment and imperial England relevant to the necessity for long ocean voyages and the need for safe and directed passage.

From this running start, those disciplines had an important and continuous place in the subsequent history of Australia. Navigational sciences shared with the other important division of Enlightenment science, "Natural History" (introduced by Joseph Banks), the guiding fabric for scientific activity in colonial Australia, relevant to the solution of a constant array of practical problems. All of this is consistent with the idea that everyone was a scientist driven by the need to use the "freedom of one's own intelligence" to achieve outcomes – be they mechanical shears for sheep or redirection of east flowing rivers in the Snowy Mountains Scheme. The point is that mechanical inventions and the construction of infrastructure involved identification of problems, and their solution using the scientific method. Astronomy was a lightning rod from the time James Cook observed the transit of Venus across the Sun in 1769, to the discovery of radiophysics in the aftermath of World War II by physicists using long wavelength images to explore the Sun and the furthest reaches of the universe. It was the "new physics"

of nuclear science that catalysed a "shuttle academia" between Australia and England (especially through the Cavendish Laboratory in Cambridge). The Cavendish had a second role, providing a blueprint which identified the value of protected research space for quality research especially for the "big questions" needing solutions. Mark Oliphant's research with Ernest Rutherford at the Cavendish, and later at Birmingham, was critical to the establishment of the postgraduate Australian National University and its School of Physics in post-World War II Canberra, as a flagship and model for the subsequent academic development of physics and astronomy in Australia.

Initially the momentum of natural philosophy had been maintained by the surveying skills of Thomas Mitchell with his monumental trigonometric survey of the Nineteen Counties tied to the Parramatta Observatory, and the quality charts of northwest Australia and the Inner Passage of the Great Barrier Reef by Phillip Parker King. Then followed observational astronomy by colonial observatories and talented amateurs, especially John Tebbutt the comet discoverer, who reminded the world of the value of exploring southern skies, and Lawrence Hargrave who discovered the principles of flight that proved invaluable to the Wright brothers' heavier than air flight in 1903. These and other named and recognised engineers and scientists are only the tip of the iceberg of talented men and women, whose contributions in mechanics, survey and astronomy made a difference to the growth of a nation, and the quality of life of its citizens.

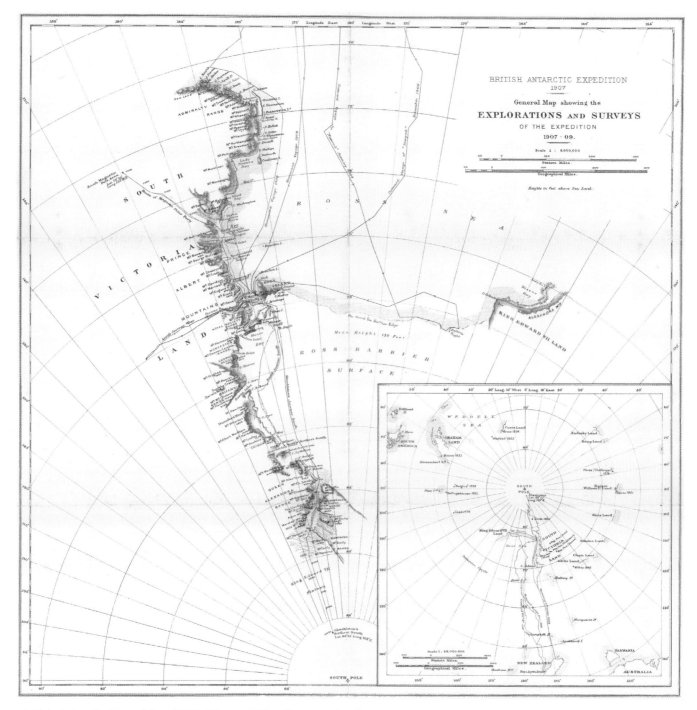

British Antarctic Expedition 1907, "General Map Showing the Explorations and Surveys" in *The Heart of the Antarctic*, E. Shackleton (1909)

This map records David's and Mawson's route to the South Magnetic Pole, completing the map of terrestrial magnetism begun by Edmund Halley. Data collected by Australians were crucial in acceptance of the Australian Antarctic Territory under the 1959 Antarctic Treaty. Australia's contribution began with Tasmanian physicist Louis Bernacci, on Borchgrevink's 1898 expedition. Many Australians followed, including Douglas Mawson, foremost Antarctic scientist of all time. Gordon Hayes describes Mawson's 1911–14 expedition as "the greatest and most consummate" Antarctic expedition. It was the individual who stood out until 1957—International Geophysical Year, which involved twelve nations using forty bases, and changed how Antarctic science was conducted. Data collected on the 1911 expedition set a benchmark when published in twenty two volumes. Highlighting the value of Antarctic science, Mawson and Madigan noted geological resemblances with the opposing Australian coast, while Wegener in Germany was publishing continental drift theory.

The End Game:
A Fabric For Science

The idea for this book came from unease about where science in Australia sits in terms of its value and of its relationship with those involved, and with those who benefit. This led to the idea that the quest to understand today's complicated world of science in its Australian context may gain clarity from examination of what went before, and from discovering the continuities between periods, and considering whether lessons for the present can be learnt from the past. In the event, it is the remarkable people encountered in the course of coming to grips with the mechanics of Australian science in Colonial times and the first half century of Federation who have become front and back the story of this book. They are the heroes of Australian science, though few of them are known today, and usually only if they appear on a stamp or banknote. So the scope of this book is the story of science in Australia to World War II and its immediate aftermath. Looking at the past and today's world, the 1940s appear as a watershed.

The Long Enlightenment was completed while Australia – indeed the world – faced the greatest health challenge since the Spanish influenza pandemic in 1919. A theme of the book is that while Australia has built a respected position within the international community through its successful response to challenges within a science-based framework, evidence suggests that in the recent past, it was losing its cutting edge. Perhaps in future, historians will look back on the Covid-19 crisis as another watershed.

Spanish influenza hit an Australia on its knees, after four years of an enervating world war and a loss of 62,000 lives – 2½% of its male population. Yet lessons had been learnt from earlier epidemics of smallpox, bubonic plague and influenza.

Though Australia's death toll of 12,000 in 1919 was devastating to a nation of five million, the influenza mortality rate was half that of New Zealand and England, thanks to countermeasures taken on scientific advice. A nascent Commonwealth Serum Laboratories

(CSL) formed in 1916 to ensure local control of essential vaccines, sera and anti-toxins – and working closely with the equally young Walter and Eliza Hall Institute – developed what was called a vaccine, using bacteria collected in infected airways. They made it available immediately to the public in a crisis situation. Those inoculated had a 40% reduction in mortality, contributing to the national mortality differences noted above.

The medical orientation of the time, set by Louis Pasteur and Robert Koch, was that bacteria cause infections and antibody protects against infection. This was before the idea of virus infection had become established. We now understand that "polybacterial vaccines" stimulate mucosal protection, which acts as a defence against viral infection of the respiratory system. An oral form of the CSL 1919 injected "vaccine" is currently being trialled for Covid-19 prevention or

"Butterflies", Plate xx, in *Australian Lepidoptera*, A. Scott (1864)

Women were already contributing to Australian science in supporting roles in an early period when social mores prevented their recognition as scientists. *Australian Lepidoptera*, written by their father, highlights the scientific and artistic talents of Harriet and Helena Scott.

modification. The important point is that a safe product, logical within the scientific framework of 1919, was rapidly made available to the public without political or bureaucratic interference – and it saved lives.

CSL established itself as responsive to medical need, becoming a leader in the international effort to identify, and rapidly respond to, mutations in influenza requiring vaccine adjustment. Australia's response to the Spanish influenza experience was evident in responses from both government and scientific institutions, each through extraordinary men. Leadership on the federal administrative front came from John Cumpston, who initiated the formation in 1921 of an integrated Department of Health, followed by a Federal Health Council in 1926 (precursor to the National Health and Medical Research Council, formed in 1936). On the science front, the Walter and Eliza Hall Institute aligned its agenda to infection and the new virology. There, Frank Macfarlane Burnet led the world in influenza research at every level – from the biology and structure of the virus to initiatives that would prove central to the production of a true vaccine in the early 1940s.

A second epidemic in 1919 – called Australia X Disease, now known as Murray Valley Encephalitis – was detected for the first time in south east Australia, with 280 diagnosed and a mortality of 68%. What was then a crisis in its own right, was lost in the confusion of influenza. Cleland (in Sydney) and Breinl (in Townsville) described the clinical and laboratory features – with Breinl transmitting the disease to laboratory animals (the first such transmission of an arbovirus) followed by French and Burnet at the Walter and Eliza Hall Institute in 1950 isolating and characterising the virus using methodology Burnet had developed for influenza research. This enabled the development of serological markers, which made it possible to trace carriage of the virus by water birds to flooded regions in the Murray Valley from reservoirs in northwest Australia. A remarkable and complete analysis of this fearful disease led to the introduction of "sentinel chickens" – domestic chickens that were monitored to detect any re-appearance of the deadly virus (in the same way as canaries are used in mine shafts to detect toxic gases). Sero-epidemiological studies showed that

this virus rarely caused clinical disease (manifest in signs and symptoms), while asymptomatic infections were common.

The point is that the responses to these great health challenges of 1919 were targeted and science based, supported at all levels, focussed on outcomes and led by outstanding people. Can we expect similar progress from the current coronavirus crisis? Or have decision making and policy become consumed by bureaucratic and post-modern social theories encroaching on academia and society, that reject Western science as fantasy – leaving little room for the innovative and creative individual, who was the centre pin of problem solving in an Australian context?

Briefly, lets return to the two questions asked in the inroduction: has science in Australia before World War II significant value? And if so, does it connect with the "science industry" of today? I would suggest there are four main messages about the character and value of early Australian science to be gleaned from a historical review, plus a further couple of secondary significance. The first was that science following James Cook and Joseph Banks in Australia was framed by a profound underpinning influence, the Enlightenment. The history of Western science teaches the importance of an overarching principal for the integrity of any component of science. The laws of motion and universal gravity defined by Isaac Newton in his *Principia* published in 1687 set the pace for such principals, establishing physics as a way of understanding the cosmos in precise mathematical equations. Darwin with evolution by natural selection did this for biology, Mendeleev's periodic table with a classification of elements based on atomic mass gave a credibility to chemistry and with plate tectonics geology gained a unifying theme.

The tenor of Australian science in Colonial times was an effect of the influence of the Enlightenment, a movement that combined the empiricism of Western science with a focus on an individual's potential which emanated from the reformation. In England, individualism and initiative were buffered by tradition and privilege – colonial Australia was a very different environment. Colonists "self selected" from many backgrounds (some illegal), all prepared to have a go,

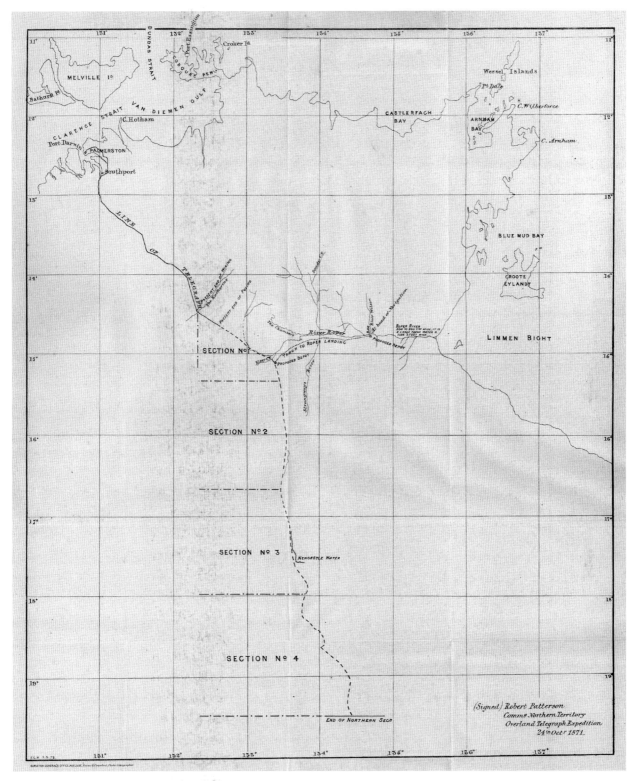

"**Northern Section of Overland Telegraph**" (1871) in *Reports on Overland Telegraph Construction* (1872), **South Australian Parliamentary Paper (83)**

Communication with ideas and discussions in Europe was basic to the more formal development of science in Australia. Completion of the Overland Telegraph between Darwin and Port Augusta in 1872 was a milestone. The cities and towns of Australia, reliant till then on sailing ships for overseas communications, instantly became part of a global network. The great engineering feat, requiring 30,000 iron poles and repeater stations every 250 km, was completed under the control of the astronomer Charles Todd.

to establish themselves, using scientific method to meet and overcome the myriad challenges they met on a near daily basis in their remote new land. The value of science within this frame was recognised by early governors, by explorers and even by Francois Péron (a scientist in the expedition led by Baudin) – though a growing and increasingly independent population would disagree with the official vision couched in imperial terms as a "Great Britain independent of all foreign nations."

The first message is that the Enlightenment gave belief and focus to the average settler, enabling the development of a society very different to the one they came from. And the Enlightenment introduced the science disciplines of James Cook and Joseph Banks – navigation science with astronomy, survey and geomagnetism, and natural history. These disciplines would continue to be a dominant component of science in their various guises. For example, in Australia today more than seventy per cent of science publications are within these disciplines. The excitement and relevance of the Enlightenment for colonial New South Wales were not lost on the ex-medical man, recent businessman and senior society figure, Alexander Berry. In his review of geology published in Barron Field's *Geographical Memoirs* in 1825, he mused on the good fortune that "colonisation was deferred until the present time when the sum of knowledge both moral and physical, is so extended that these attempts (at discovery) may be made upon just and rational principals."

The second message is that thanks to these impacts of the Enlightenment, everyone was a scientist. People analysed and solved problems using empiricism and experiment. "The advance of science," in Phillip Parker King's conception, was "common to every enlightened mind". Historians see science as part of the daily life of colonists who applied it to problem solving and growth. Thus, Julie McIntyre researching "Settlers in the Empire of Science" took King's son, James, and John Macarthur as examples of scientific farmers in mid-19th century New South Wales. John Gascoigne's "The Enlightenment and the Origins of European Australia" concludes that "scientifically-based improvement could change the landscape." The 1868 rules of the Agriculture Society of New South Wales emphasise "the enlightened

combination of Practise and Science". The advent of Mechanics' Institutes for the technical education of artisans in Edinburgh in 1821, inspired by the Scottish Enlightenment, was followed within six years by similar institutions in Australia – in Hobart (1827) and Sydney (1833). They presented talks and instruction, as well as science education. By 1890 there were over 2,100 mechanics' institutes, also known as schools of arts, in Australia. They provided libraries and courses of high interest to the community. Science was embedded within the community.

The third message is that science in early Australia was more than a system of thought – it was a means to survival and economic growth, a mechanism to turn a problem into a profit. The grandiose idea of Governor Brisbane that science would "stand the empire alone", was lost on the colonists, whose interests were strictly personal. The point is that nearly all scientific endeavours had a commercial component. Even scientific studies on natural history were aimed at the bigger picture of imperial power with its economic implications. In agriculture and animal husbandry, increasingly reliable and widespread disease free crops, and larger quality fleeces, were part of the harvest of science. Behind every initiative was a scientific plan and most importantly, a remarkable person making it work. From John Macarthur to Arthur Turner in the selection and health of the merino sheep, to William Farrer, Arthur Prescott and John Ridley with their ground breaking contributions to developing the export wheat industry. Economic geology and mining were always a driving force in earth sciences, from the time of Rev. William Clarke (who first discovered gold) through to Edgeworth David, Douglas Mawson and Edwin Hills. Physics and the closely associated disciplines of astronomy and survey have traditionally been linked to the widest range of infrastructure including communication, navigation and transport infrastructure.

Contributors include Hargrave who pioneered heavier than air flight, Mark Oliphant with radar and nuclear power, and Joseph Pawsey pioneering the radioastronomy studies which made possible Wi-fi. Similar lists of names in medicine, biology and chemistry underpin modern medicine, biotechnology and the chemical world. The link between pre-World War II science and economic advance is intricate and comprehensive, but always anticipated.

The fourth message concerns the role of British academia. In the timeframe examined here, the ever changing relationship between Australian and British science programmes was always critical to the development of science in Australia. The first of three broad periods, the Banks era, began with Cook's and Banks' discovery of the east coast in 1770, and ended with the death of Banks in 1820. Banks dominated scientific study of Australia with the clear view of New South Wales as a feeder colony for British scientists. "Collectors" and scientists such as Robert Brown put Australian natural history out for all to see, but with little thought about establishing programmes based in Australia.

The second period of British-Australian scientific relationship was a watershed: European scientists began to call Australia home, contributing to establishing bases for research and spending more time in Australia. This period from 1820 to about 1870 included scientists involved in natural history such as Cunningham, Clarke, and von Mueller, and astronomers including Rümker. Outstanding survey work by Phillip Parker King and Thomas Mitchell brought international attention for Australian based scientists. Late in this period, academics were recruited into the new universities while other professionals migrated for reasons of opportunity or health, increasing the pool of individuals seriously focussed on science. The trickle of Australians getting professional training in Europe, mainly in medicine, slowed as places became available in local universities. The prominent names amongst scientists in this period were recent immigrants from Europe, but Australian born and trained scientists were beginning to appear in the 1870s. Harry Allen, for example, the first Professor of Pathology in Melbourne (1882), was an outstanding and influential academic, born and trained in Australia.

The third period was important for Britain as well as Australia. It was characterised by an academic shuttle system bringing young British academics of great talent but limited professional prospects to take advantage of opportunities in science or medicine in the three Australian colonial universities, where they established strong international reputations. They maintained close connections with British colleagues, especially in Oxford, Cambridge and London. Talented Australian graduates, through these connections and aided by British scholarships for the best and brightest, had an opportunity to gain research training and PhDs – neither of which was available until the mid-20th century in Australia except at a few research institutes such as the Walter and Eliza Hall Institute. Some British academics would return from Australia to Britain to influential research positions available to them after an Australian experience. These shifts resulted in the development of "Australia friendly" centres in Britain with a track record of attracting the best of young Australian scientists, a situation that continued through the first half of the 20th century. Some outstanding Australians would stay on in Britain, taking senior positions where they also became centres of Australian interest, including Frank Debenham at Cambridge, Henry Harris and Howard Florey at Oxford, Mark Oliphant at Birmingham, Gordon Cameron and John Cornforth in London – to name a few. Two leading examples illustrate the influence of Anglo-Australian relationships on patterns of science.

First, the Cavendish Institute in Cambridge with its impact on physics and indeed many would say, on organised science in Australia. In the mid-19th century scientific research was done off campus, with experiments in colleges and private premises. The flurry of discoveries at this time by the likes of William Thomson (Lord Kelvin) and James Maxwell led Lord Cavendish in 1871 to finance a dedicated research facility in Cambridge with Maxwell as director – essentially the beginning in Britain of institutional research and a direction of great subsequent importance for Australia. Thomson's discovery of electrons, and the detection of the atomic nucleus by his successor Rutherford from New Zealand, established the Cavendish as the centre of physics research and a beacon for bright young physicists. Rutherford was head of the Cavendish from 1918. It became a mecca for young Australian physicists, beginning with Mark Oliphant after Oliphant met Rutherford in Adelaide.

Australian academia began under Oxbridge influence. Sydney University's coat of arms included the emblems of Cambridge and Oxford, and most of the foundation professors at Sydney, Melbourne and Adelaide universities came from Cambridge or Oxford. The Cavendish was central to this early influence. Horace Lamb, the foundation professor of Mathematics at Adelaide in 1875, was a Cambridge graduate taught by Maxwell, but it was Thomson who had the most influence, through influential second wave appointments in physics from the Cavendish in the 1880s – Threlfall in Sydney, Laby in Melbourne and Bragg in Adelaide. Bragg's son Lawrence, born in Adelaide, would succeed Rutherford as Director of the Cavendish in 1938. Lawrence Bragg would provide leadership also in the Royal Institution, introducing x-ray diffraction technology to analyse structure of proteins and DNA.

The tradition of the best physics graduates from Australian universities going to England for postgraduate training began when William Bragg sent his two highest achieving students to work at the Cavendish with Thomson. The flow became a torrent between the Wars when Laby was Professor of Physics at Melbourne University. Of the twenty one awarded 1851 Exhibition Scholarships in Melbourne, twelve were graduates from Laby's department who went to Cambridge. Oliphant continued the Australian procession in search of a future in physics when he went from the Cavendish to Birmingham. Australian colleagues in his laboratory there included John Gooden (who returned to Adelaide), and Jack Blamey, Len Hibbard and Wibs Smith, who all moved with Oliphant to the new national university in Canberra. With so many young Australian academics influenced by their postgraduate experiences in Cambridge, the impact of the research atmosphere in the Cavendish on late 20th century Australian university culture isn't surprising. In the early 1980s – when Australian universities were struggling for funds and identity, the Professor of Physics at La Trobe University in Melbourne, John Jenkins perhaps with prescience, wrote of a malaise in Australian universities in part due to the Cavendish tradition, characterised by a distancing of physics from a strong mathematical base, a neglect of the new American focus on modern equipment, and a lack of interest in anything applied or practical.

The second example was Charles Martin and the Lister Institute in London. Martin, from a modest background, worked his way through university with a medical degree at St Thomas' Hospital, London, then took a lectureship in medicine at Sydney University in 1891. Few were more impressive, nor better adapted to Australian society than Martin. First in Sydney where his work on snake venom and comparative anatomy and physiology of marsupials attracted international attention, then as professor in Melbourne. He returned to England to take the directorship of the Lister Institute in London in 1903, which he held until retiring in 1930. Here he established a world leading centre in host-parasite medicine, leading teams to tackle infections challenging the profession of the day – typhus, the plague, and typhoid-related disease. He maintained his close connections with key biomedical researchers in Australia for the rest of his life. In World War I he joined the Australian Army Medical Corps, again making important contacts. After retiring from the Lister, he returned to Australia to work with the CSIR for several years. It is hardly surprising that many Australians seeking experience in biomedical research sought Martin and the Lister – including Howard Florey, Macfarlane Burnet, Roy Cameron, Charles Kellaway and Hamilton Fairley, a who's who of medical research in early 20th century Australia, centred around the theme of host-parasite relationships. Indeed, together with neuro-physiology, the science of relationships between man and parasite (infection) has dominated medical research, culminating in the discovery by Robin Warren and Barry Marshall of *Helicobacter pylori* as the cause of peptic ulcer, a disease that at that time dominated clinical medicine.

A fifth message is the dominance of targetted science. Practically all science in Australia before 1950 was targetted to a problem of concern. Today we recognise this as applied science, and despite its reliance on scientific method, and the importance of outcomes, many academics are dismissive. This was understood by C.P. Snow in 1965 when he wrote "scientists have by and large been dim-witted about engineers and applied science. They couldn't get interested. They wouldn't recognise that many of the problems were as intellectually exciting as pure problems . . . applied science (in their view) was an occupation for second-rate minds." Indeed, their negative view largely accounts for the poor recognition today of the part played by scientific method in handling the problems and challenges in the 150 years following white settlement.

The impact of war on Australian science presents a sixth message. Much has been written and will not be repeated, but several observations with respect to science in postwar years should be made. An impact of World War I was that the Commonwealth Government recognised both the practical value of science and the importance of public control of scientific industries relevant to providing essential services in an isolated country. The development of Commonwealth Serum Laboratories as a government entity to provide essential vaccines (and later antibiotics and blood products), and the CSIR to provide protection for primary industries, are conspicuous examples. A remarkable thing was that some individuals shared this concern over isolation from essential resources. John Peake, recovering in England from wounds sustained on the Western Front, retrained in chemistry with the specific objective of starting a research-based company in Sydney to supply essential chemicals. Timbrol was born. War had a catalytic effect of interaction of individuals and organisations at many levels – with outcomes ranging from Basedow's expedition into Northwest Australia looking for a source of tungsten, to the enlistment of Charles Martin. Martin, who was back in England at the Lister Institute when war broke out, joined in the medical section of the Australian Imperial Force, where he would influence young scientists such as Hamilton Fairley, consolidating the Lister's link with Australian biomedical science.

World War II had an undeniable impact and again much has been written. Australian scientists such as Florey and Oliphant played essential roles in the ground breaking developments of penicillin, radar and nuclear fusion and the harnessing of energy in the Manhattan Project in America. In Australia, university science departments and government research organisations switched focus to whatever was needed for the war effort. Outcomes and impact ranged from the CSIR's work on radar – which led to postwar leadership in radiophysics – to improved gas masks and insecticides, and the grinding of gun sights. Whatever it took.

This period was a watershed: long-standing and often somewhat worn out senior academics retired and were replaced by a new breed of scientists with European training. These were either Australians returning from post-graduate training in Britain or recruits from amongst the massive post-war immigration programme which by 1950 had brought 200,000 migrants from Europe. They began a transformation of Australian society, strengthened by one million migrants per decade for the next forty years, many with technical and science training. This period resembled the gold rush era of 1851–60 when 50,000 people arrived each year (including many discussed in this book who took leading roles in science). The CSIR became the CSIRO in 1949 and expanded from its initial focus on primary industries, to include coal technology, wool textiles and metallurgy, as the need for manufacturing was recognised. The National Health and Medical Research Council (NH&MRC) signalled a practical entry into health by supervising the introduction of childhood vaccine programmes including oral poliomyelitis vaccination.

Perhaps the most specific outcome of World War II for science was the government initiative to develop a national postgraduate research university in Canberra (the Australian National University) to focus on the best of Australian science, in order to compete in the international world of science on a level playing field. The two UK-resident Australian scientists who had burst into recognition because of their war efforts, Howard Florey and Mark Oliphant, became scientific advisors and an extraordinarily talented group of scientists was recruited to lead the research schools, including Oliphant as head of Physical Sciences and Frank Fenner as head of Medical Research.

Conspicuous in this overview of scientific achievement before 1950 is the dearth of women scientists. A review of the authorship of papers presented to the Royal Society of New South Wales reveals that between 1920 and 1940 male scientists with academic positions dominate. Over the twenty years, careers and patterns of research can be traced. Many authors became heads of department, though few demonstrated significant innovation. Arthur Penfold (1890–1980), phytochemist and museum director, painstakingly worked his way through most Australian native plants, defining the chemistry of the volatile oils, presenting results in two or three papers every year, but never in association with a female co-author. By contrast, the organic chemist Francis Lion, whose studies in structural aspects of complex chelates were years ahead of work elsewhere in the world, averaged six to seven presentations each year usually co-written with female postgraduate scientists. One of them, the brilliant Rita Harradence, was awarded an 1851 Exhibition Scholarship to work with Robert Robinson at Oxford on sterol biochemistry. She was later a close and long term collaborator with her husband, the Nobel Prize winner, John Cornforth.

Women in Australian Science

Women were under-represented in an age long before today's occupation with gender equality. Through the 1920s, only one to two per cent of communications in the *Journal of the Royal Society* (NSW) were authored by women. In this era when women were not recognised as potential leaders in science, two of them made significant inroads into male dominated academia in the field of earth science – Maria Bentivoglio and Ida Brown.

Maria Bentivoglio was the first woman to have an 1851 Exhibition Scholarship to Oxford University. Her main interest was crystallography of minerals. She returned to Sydney University, where following the departure of Thomas Griffith Taylor in 1926, she became head of the Department of Geography. She later pursued a career in school education and industry in the U.S.A.

Ida Brown (1900–76) had an impressive but poorly rewarded career at Sydney University. She was inspired by Edgeworth David, gaining a local Linnean-Macleay Scholarship to complete fieldwork and petrology (hard rock geology) on the south coast of New South Wales, for which she was awarded a DSc. Her research shifted to soft-rock stratigraphy in the Yass area, then to Palaeozoic fossil studies. In an age where opportunities were generally closed to married women, they both chose careers instead of marriage, until at the age of fifty, Ida Brown married her boss, and was obliged to retire.

The pattern began to change in the 1930s. Before the War women presented four to five per cent of the papers, but in the War years one third were by women. The increase in academic opportunities for women, while most young men were called into the services, paralleled opportunities created by the war effort such as employment of women physicists by the CSIR in the radar project. Yet the restrictions imposed by male perceptions, and retirement enforced by marriage and pregnancy, continued. Only three of those women who presented papers at the Royal Society continued academic and research careers. Rita Harradence, Germaine Joplin (1903–89) and Dorothy Hill (1907–97).

Germaine Joplin got the University Medal and a BSc at Sydney University in 1930 and a PhD at Cambridge. She returned to Sydney where she completed three important compilations of analytical data on Australian rocks, writing the comprehensive Petrology of Australian Rocks published in 1964. She was awarded a DSc in 1950, but had no permanent position until employed in 1952 in the Department of Geophysics at the new National University in Canberra.

Dorothy Hill, a Brisbane graduate in geology, would become the first female professor at an Australian university and the first female President of the Australian Academy of Science. She obtained a Queensland University Travelling Scholarship to complete a PhD at the Sedgwick Museum in Cambridge in 1930 where she pursued an interest in fossil coral. She developed an understanding of morphological relationships in fossil coral, which she developed into a sophisticated system of taxonomy enabling fine tuning of stratigraphic palaeontology. This she built on in subsequent contributions to the geological mapping of Queensland. She returned to Brisbane in 1937 on a CSIR grant, joining her old department. Working at a time when academic rigor was lacking, Dorothy Hill was particularly concerned to reverse the prevailing torpor and to encourage research by students. Her own work used fossil corals to sort correlations in palaeozoic rocks.

She took advantage of proximity to the Great Barrier Reef while expanding knowledge of stratigraphy in eastern Australia. She brought wide knowledge and understanding to both academic and economic geography.

Both Joplin and Hill had to give their lives to their work. They became respected advisors and consultants and never married. While an unexpected outcome of World War II was opportunity for some women scientists, retention and promotion within career pathways would take longer. The challenge for young women in science can be exemplified by looking at the career of Ruby Payne-Scott (1912–81). She had a brilliant academic record, obtaining first class honours in maths and physics at Sydney University in 1933, with prizes in both disciplines. After various positions including teaching and as a librarian, she became involved in receiver design at Amalgamated Wireless (Australasia). Her most important contribution was, with Joseph Pawsey, to develop mathematical approaches to analyse emissions from the Sun of an infinite series of waveforms (or Fourier components), enabling computing by Fourier transformation. Published in 1946, this work became the mathematical foundation of radioastronomy. Ruby Payne-Scott was a woman of strong views and social constructs. She had to resign in 1950 following her marriage. She worked as a science teacher but never returned to astrophysics.

Then and Now

It can be concluded that in Australia science in the first 150 years or so after colonisation not only shared in the extraordinary growth of the nation but was significantly responsible for it. The immediate post-World War II period became a watershed because the parameters around the very nature of science changed. Whereas immigration boosts in the past had related to wool and gold, extending existing Western culture and reinforcing interaction with Britain, migration followed a very different pattern after the War. A population less connected to inherited British traditions would evolve in the second half of the 20th century. The British way of science would be challenged by European and increasingly by American influences. Science was partitioned into university departments. The role of ordinary people diminished as the problems that needed solution shifted into bureaucratic tangles. Applied research suffocated amongst the calls for basic science. Losing targets meant a blurring of focus. Large

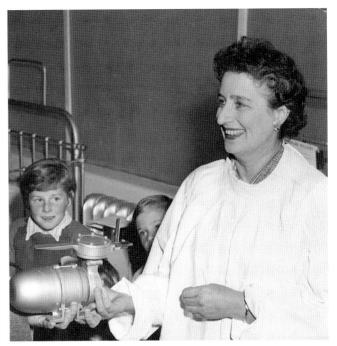

Charlotte Anderson

Charlotte Anderson was a pioneer in paediatric gastroenterology research, making important discoveries related to coeliac disease and cystic fibrosis, including refining diagnostic tests and developing gluten free diets. Later in her career she was appointed the first female professor of paediatrics in Britain.

discipline departments appeared across the academic spectrum, with remarkable similarities to one another (and to the international agenda) and attempting to be all things to all people, be it in physics or biomedical sciences. Despite the massive increase in numbers of science graduates, this diverse proliferation diluted effort.

An idea of the expansion of science activity and a change away from the earlier balance where applied science was the main game, is evident in looking at government funding. The CSIR began in 1926 with a grant of £250 to fund the eradication of prickly pear. Twenty years later the annual funding for the CSIR was fifty times the total for research in universities. The initial grant to the NH&MRC in 1937 was £37,000, which funded fifteen fellowships and thirty one of sixty three project applications. The average grant was £730, with the largest sums awarded to future Nobel laureates, Macfarlane Burnet and John Eccles. The funding was solely for medical research. In 1946 to remedy the funding deficiency in other areas the Australian Research Committee began a funding programme. It was replaced by the Australian Research Council in 1966.

In 1966, 406 successful applicants were funded from a pool of $4 million, at an average of $1,000 per grant. The sheer growth of tertiary education is evident in the numbers of undergraduates. In the late 1850s there were 140 students in the Universities of Sydney and Melbourne. By World War II about 10,000 were attending six small state universities. In 2015 there were forty three universities in Australia, 1.4 million students and 118,000 staff. About twenty five per cent of all high school graduates were being admitted to university! Looking at PhD awards as an index of research activity and a reflection of research infrastructure within universities is telling. The PhD is a research degree that was not available in Australia until 1948, and currently more than 10,000 are awarded by Australian universities each year! Funding for research by the federal government has risen to the order of $3 billion each year, split fairly equally between the CSIRO, the ARC, and the NH&MRC. The number of grant applications has increased 500-fold over fifty years, yet the success rate is consistently around fifteen per cent.

Superficially this reads as a success story, though the low success rate for grant applications and the enormous number of PhDs awarded suggest closer examination. Publication record analysis placed Australian research at or above world standards in sixty per cent of designated research disciplines with sixteen new inventions on the patent register per university and in excess of 430,000 research papers published each year. It would appear that research growth comfortably fits with economic growth over the last seventy years and that Australia is punching above its weight on the international stage. That fits with a research contribution of 2.8% of world total when Australia's economy is at 1.7% of world GDP.

However, the contribution rate of 2.8% when adjusted is in fact lower than that of countries with comparable OECD rating. When more penetrating indicators of research quality are examined, the standing of Australia is less impressive, even trending downwards. Perhaps this does not surprise when for some years the comparative standards on a world playing field of performance of Australian students in maths and science are falling. The citation index measures the value peers attach to published research. On this basis, twenty five years ago Australia was ranked 7th. Recently it has fallen, fluctuating between 9th and 16th. These figures apply across various research fields. In eighteen science segments, Australia rates above a European average in only five, which compares to the UK which ranks above average in all eighteen!

Perhaps a pointer to contemporary Australian science, is the relatively few discoveries related to big questions. For example, of the fourteen Australian Nobel Prize winners in science, approximately two thirds were recognised for work done before the 1950s. No prizes were awarded to scientists in undergraduate universities, and all but one were won by scientists working in a dedicated research spaces. The exception was the award for discovery of *Helicobacter pylori*, the bacteria colonising the stomach that cause peptic ulcers. That work is reminiscent of the discoveries by individual scientists in pre-War Australia, who identified a problem and used a logical and simple scientific approach to solve it. Barry Marshall and Robin Warren were not career scientists, but clinicians who used their own intelligence to solve the greatest medical problem of the day (and were dismissed by their academic colleagues).

In modern times there may be an image of pressure on scientists to achieve practical outcomes. Indeed, there have been outstanding successes: positive airways pressure to treat sleep apnoea by Collin Sullivan in 1981, the bionic ear by Graeme Clark in 1978, the Black Box Flight Recorder by David Warren in the 1950s, cardiac pacemakers by Geoff Wickam in 1963, and effective eradication of *Helicobacter pylori* and cure of peptic ulcers by Tom Borody in 1987. The list is long and impressive, but most discoveries were made by individuals with a commitment and passion to resolve a problem, and few came from university-based academics. Several of these initiatives have developed into multinational companies such as Telectronics, Cochlear and ResMed, and many have not.

A comparison of listed biotechnology companies in Australia compared to the USA, France, Germany and the United Kingdom, is of interest and reflects a deficiency in the process of translational research in Australia. In a 2015 survey by the business management company PWC, Australia with thirty seven listed biotechnology companies was second only to the USA (with 315). Significantly fewer were listed in the UK and Western Europe, with about half the number of companies, but two to three times the market capitalisation. While 51% of Australian listed companies had no revenue, in Europe this number was about 25%, and in America only 15%. There were too many companies with an average of $6–$7 million market capitalisation (compared with the USA's $2 billion and the UK's of $340 million). These are often one product companies, with insufficient funds to survive the costly process of bringing the product to market. Few biotechnology companies with a market capitalisation of less than $10 million survive.

The purpose of this book shifted away from the initial stimulus which was an unease about directions being taken in contemporary science, to explore the use of scientific method by remarkable people to shape a nation. Returning to the linkage between science in Australia before World War II and science since the War and whether there are lessons of value to be learnt from early experience, six characteristics of the earlier period can be contrasted with current patterns, particularly within universities. Science has become concentrated there, as is evidenced by the number of postgraduate research scientists and the volume of government funding, where in excess of sixty per cent of research money funds university programmes. Indeed, universities have become the benchmarks of modern science. The politicisation of the CSIRO and loss of its unifying and guiding vision have blunted applied science at a national level. Alan Finkel, Australia's Chief scientist has commented that the absence of an underpinning principal applies more broadly in today's world of science.

The first of the six characteristics is that before 1950 science occurred within an Enlightenment framework which gave a foundation and energy to Australians who saw the glass half full with no limits to what they could achieve. The use of the scientific method was second nature as was the idea of using one's own intelligence

without restrictions. Perhaps this is the fundamental difference between earlier and modern times. It is seen as such by John Cornall, Emeritus Professor of Sociology at La Trobe University, who describes a disintegration of Australian universities in which fifty five per cent of staff are bureaucrats, and universities lose "their unifying and guiding vision" while academics fail to "resist bureaucratisation". Science is compartmentalised and less a part of society and role models are uncommon. The Enlightenment and its influence persist to the extent that individuals, generally outside universities, continue to make amazing strides, but within the halls of academia the light at best is dim. The very idea of Western culture of which the Enlightenment was a shining part, is anathema to the university of today.

How does academia recover an underpinning principal? Without one, research and science become a commodity, losing focus with research agendas all over the place. Currently much of the research is designed and rewarded on process rather than ideas – a dangerous place to be. Process, including slavish adherence to analytic dogma such as Popper's holy cow of falsification, and obligatory identification of compulsory objectives prior to study (imposed by drug regulatory authorities), denies the growth of knowledge, discourages the search for new ideas, and confounds the value of outcomes. (The influential philosopher, Karl Popper proposed that "falsifiability" be taken as the essential basis for testing scientific validity, rather than substantiating the truth of propositions.)

Method can deny progress. In the real world of grants and the finance of science, insistence on "process-correct" science irrespective of the question being asked trumps attempts to model data or identify something novel and important. Single data points not fitting the hypothesis need to be understood but not used to negate the idea. So it is that much low importance research, beautifully done and eminently publishable, floods university research agendas and PhD projects, perpetuating the risk of mediocrity. Fifty per cent of current research publications have no peer citations, an accepted index of research value. Perhaps a sea change is needed, in that the idea of a single great truth like those which characterised the age of Newton, is appropriate. In an Australian context that is happening with some success in biology around the theme of host-parasite relationships and in astronomy with its drive to examine defining events in the early stages of an expanding universe. An important step is to rekindle the excitement and enthusiasm for science and the scientific method amongst the public at large, which was so much a part of earlier Australian society with its drip-up impact on a research culture. This is about leadership, role models and education. Can we transform a politically correct motif into a "science (reality) motif" in our society? In Australian universities today up to fifty per cent of students are full fee paying, a situation that conflicts with research values and priorities, making difficult the recognition of any "underpinning principle". It is often suggested that "innovation" fills that bill – as a term reflecting converting ideas to outcomes, with a direct link to the economy and quality of living. In an attempt to rank countries based on measurements in sixteen activities, the 2019 Bloomberg Innovation Index has Australia at 19th. In 2015 it was 15th. Number one in both years was South Korea.

The second characteristic is the importance of the individual in identifying, responding to and resolving specific problems. Many examples have been discussed in this book, and in this respect individuals continue to make important discoveries, usually from an independent background. Classically individuals took on a specific task, usually of a commercial nature. How do you prevent disease in sheep? How can wheat be protected from rust? How do you process sulphide ores? Today in universities we have teams but few towering figures who stand out and lead or who are associated with important questions. We choose international projects and while often adding important supplementary data, rarely come up with ideas or solutions that make significant impact.

A more fruitful approach might be for groups to form around a leader with a central idea, such as the malaria vaccine group begun by Michael Alpers, Director of the PNG Institute of Medical Research. He identified an aim, discovering a malaria vaccine, and included several Australian Institutes and university departments. Much important work on malaria came out of this group and important studies continue, focussed around the PNG

Institute. Similar approaches have been organised, and government development of centres of excellence and related granting initiatives are promising, if well selected and not destroyed by bureaucracy. Belonging to such multi-organisational groups gives purpose to the member teams, providing real international exposures and experience to participating students. Government support for such clusters led by exceptional people and focussed on important questions is one initiative that can achieve important outcomes.

The third characteristic is the cycling of scientists and establishment of Australian friendly research institutions overseas, which in former times resulted from Anglo-Australian interdependence. Australian scientists in England had an influence on Australian science in advisory, collaborative, and teaching roles. After 1950, the world of Australian science gradually changed, in part because of increasing links with North American institutes and changing demographics in Australia. There was a greater focus on needs for expensive equipment and large budgets, and competition with international science particularly as practised in the USA. A thin line ran between competing with others' agendas, and evolving ideas based on new information. A case can be made for a clearer enunciation of important questions relevant to Australia and its position in the world, with strategies developed to find solutions. The Anglo-Australian shuttle involving both scientists and ideas perhaps can be reshaped for a post-Brexit era. The current innovation planned for the Cavendish Institute, linking industry and science through education and research activity, is an example of value for cooperative ventures. In the modern era we educate many Asian students who return to their countries without research and development linkages with Australia, depriving Australia of added benefits, noting that South Korea and Japan regularly top the Bloomberg Innovation Index. Consolidating associations with Asian institutions through these students could establish benefits similar to those discussed in relation to the Anglo-Australian shuttle.

The fourth characteristic is the importance to research of an economic component. Potential for economic advantage enriched early Australian science, encouraging successful outcomes and defining questions of importance. The relationship between successful science and economic gain is a constant in the story of Australian science. Today, entrepreneurial individuals continue to develop new ideas and solve problems, but most are outside traditional organisations that attract public funding, especially universities. Translational research is part of every university's mantra and with numerous patents filed, one would think research and development in Australia are in good shape. Yet figures discussed earlier show commercialisation of ideas is not done well by universities, with too many under-financed companies pursuing single product objectives, limited understanding of regulatory measures and no adequate game plan. The business sector often is more interested in early returns and is detached from any understanding of the product it would like to see produced. There is on one hand, little encouragement to develop new ideas, and on the other, inadequate ways of giving ideas a realistic chance of commercial success. Lingering British disdain for applied science, at least until a project looks like being successful, remains an obstruction.

What can be done? It is difficult because current aptitudes are complex and linked to societal change. Mentors and role models with success in translational research can help by working in further scientific research. In modern Australia translational research is disadvantaged by the lack of connection between the academic and the market. Raising money, making a development plan and dealing with regulatory agencies (and other gatekeepers) need a rethink. One idea worthy of thought would be for the CSIRO as it seeks a future, to take on the role of selecting, nurturing and developing good ideas in conjunction with the inventors. The CSIRO would become a national R and D Provider. The idea builds on the CSIRO's commercial track record and strengths across science sectors. Searching for how best to adapt to the future, the CSIRO has decades of experience collaborating with universities and commerce. In a sense taking full advantage of that is an exercise in "back to the future" and recovery of historic strengths.

The fifth characteristic is the breadth of science training and experience that some leading contributors used to cultivate. At the beginning of the 20th century Archibald Liversidge set three papers in final year chemistry, organic chemistry, inorganic chemistry, and philosophy and history of science. We hear cries for better school training in science and more funding for research. Yet large amounts of money directly and indirectly are available, and many are embarking on scientific research. For many there are no jobs, while others have a job only if their supervisor attracts grant funds. Since over eighty per cent of grant proposals fail, many good young scientists miss out on positions, and then find their training too narrow to do much else. Some universities have begun looking at this problem and are providing a broader training experience for science graduates entering the marketplace. Europeans have provided broader education experiences, and postgraduate training in America usually involves course work. A conclusion is that a rethink is needed in Australia on how science is taught and how the science workforce is employed. This implies a very different education and organisation of the working science community. It is interesting to note the example of South Korea – in the early 1950s at the time of the Korean War, the vocabulary of modern science did not exist in the Korean language twenty years later there was a national programme to familiarise the public with science, which included education revision and vocational schools. At every level South Korea now outpaces Australia. Unless a community understands and supports science, any focus on one aspect (e.g. teaching STEM subjects – science, technology, engineering and manufacturing) outside of its social context is likely to have limited success.

The sixth characteristic is the value of research within a dedicated research space. Individuals in Colonial times could manage their time and focus, but as organisations developed with primary responsibilities other than research, such as education and service roles, neither time nor a supportive environment was there for scientists. This was recognised by some academics and influential community members who observed the precedent value of dedicated research institutes such as the Cavendish in Cambridge and the Lister in London. Prime Minister Billy Hughes noticed a similar organisation of research to support primary industry in Germany.

In Australia medical research became a function of institutes developed in association with major hospitals, usually including responsibilities for the new pathology. The success of these institutes directly related to their success in maintaining an independence from hospital departments and the local university so they had control of the amount of pathology and teaching they took on. The most successful institute, the Walter and Eliza Hall Institute in Melbourne, was the inspiration of Professor Harry Allen who held the Melbourne University Chair of Pathology. It had no routine service function and was left to identify its research agenda without outside interference, so it could build a thematic base around study of host-parasite relationships. When it did develop a clinical unit, it was a direct extension of the Institute's focus on infection. Those less successful research institutes such as the Institute of Tropical Medicine in Townsville and the Kanematsu Memorial Institute in Sydney, were embroiled in hospital politics, constrained by expectations of pathology services, or considered as an umbrella for all hospital related research, creating bureaucratic nightmares without focus and priorities.

Successful applied research came from inspired leadership that understood the nexus between basic and applied research. The CSIR became a powerful and scientifically successful institution, making multiple scientific contributions for the direct benefit of the Australian Society. The value of uncluttered research in a dedicated space with an overarching theme and supportive interaction with like-minded scientists was critical to the best scientific outcomes by Australians between the Wars. Perhaps the crises afforded by the Covid 19 pandemic provide a time to consider strategies that review, re-align and integrate with industry, teaching and research in the post-Covid era, with contemporary challenges and opportunities.

In today's world, the value of dedicated research space is broadly recognised – the issues are more whether characteristics that marked earlier success are in play, such as quality of leadership, cohesiveness of theme, and lack of distractions. Common challenges are the recruitment of inspirational leaders with successful research careers aimed at big questions, managing stifling bureaucracy, keeping on song and establishing links with feeder institutions which turn out bright young postgraduate students.

The end game is these principle characteristics that emerge from an analysis of what made science very effective in Australia until the 1950s and how these provide a mirror against which the science of today can be measured. If indeed, as many believe, Australia is losing a cutting edge in the world of science, understanding these principals may contribute ideas that could restore direction and momentum to research activities, as well as again making science a topic of conversation in our community.

Bibliography

General Sources

Australian Dictionary of Biography (National Centre of Biography, 1966–2012).

Australian Science in the Making, ed. R.W. Home (Cambridge University Press, 1988).

Encyclopedia of Australian Science, University of Melbourne.

Gascoigne, J., *The Enlightenment and the Origins of European Australia* (Cambridge University Press, 2002).

Geographical Memoirs on New South Wales, ed. Barron Field (London: John Murray, 1825).

Gribbon, J., *History of Western Science* (Folio Society, 2006).

Historical Records of Australian Science, eds. S. Maroske and I. Rae (1966–2020).

Journal and Proceedings of the Royal Society of New South Wales, vols. 40–79 (1856–1945). Previously known as the Philosophical Society of New South Wales.

Moyal, Ann, *"A Bright and Savage Land": Scientists in Colonial Australia* (Sydney: Collins, 1998).

Proceedings of the Royal Society of Victoria (1854–1945).

The Commonwealth of Science, ed. R. MacLeod (Oxford University Press, 1988).

Wotherspoon, G., *The Sydney Mechanics' School of Arts – A History* (2013).

Does it Matter?

Attard, B., "The Economic History of Australia from 1788", (www.eh.net).

Fleming, D., "Science in Australia, Canada and the United States", in *Proceedings of the 10th International Congress of the History of Science* (1964), p. 82.

United Nations Development Programme, *Human Development Index* (2015).

Peters, M. and Besley, T., "The Royal Society, the Making of 'Science' and the Social History of Truth", *Educational Philosophy and Theory* 51 (2019), p. 227.

"UN Agency Ranks Australia 39 out of 41 countries for Quality Education", *Sydney Morning Herald* (15.06.2017).

Natural History

Brown, R., "General Remarks . . . on the Botany of Terra Australis", in *A Voyage to Terra Australis* by M. Flinders, (London, 1814), p. 533.

Darwin, Charles, *The Origin of Species by Means of Natural Selection*, 6th Edition (London: John Murray, 1873).

Morton, A., *History of Botanical Science* (London: Academic Press, 1981).

von Mueller, F., *Fragmenta Phytographica Australiae* (1862–81).

White, M.E., *The Greening of Gondwana* (Kangaroo Press, 1998).

Zoology

Berch, C., and Andrewartha, H., *The Distribution and Abundance of Animals* (1954).

Krefft, G., *The Mammals of Australia* (Sydney: Thomas Richards, 1871).

Agriculture and Pastoral Science

CSIROpedia, "Our History", (10.12.2015).

Dando-Collins, S., *Pasteur's Gambit* (Vintage Books, 2008).

Farrer, W., *The Agricultural Gazette of New South Wales* 9 (1898) pp. 131–168; 241–60.

New South Wales: its Progress and Resources (Sydney: Government Printer, 1886). Publication for the Colonial and Indian Exhibition.

Schedevin, C.B., *Shaping Science and Industry* (Allen and Unwin, 1987).

Anthropology

Radcliffe-Brown, A., *Social Organisation of Australian Tribes* (1931).

Spencer, W.B., and Gillen, F.J., *The Native Tribes of Central Australia* (1899).

Biomedical Science

Thompson, J.A., *Report of the Board of Health on a Fourth Outbreak of Plague in Sydney, 1904: with remarks on the aetiology of plague, based on its observed epidemiology* (Sydney: New South Wales Board of Health, 1905).

Dando-Collins, S., *Pasteur's Gambit* (Vintage Books, 2008).

Doyle, A.E., "A Survey of Australian Achievements in Medical Research", *Report to the National Health and Medical Research Council* (Melbourne, 1989).

Dyke, T., and Anderson W., "A History of Health and Medical Research in Australia", *Medical Journal of Australia* 201 (2014).

Morison, P., *The Martin Spirit* (Sydney: Halstead Press, 2019).

Walter and Eliza Hall Institute, "History of Walter and Eliza Hall Institute of Medical Research", (www.wehi.edu.au).

Geology

Austin, J.B., *The Mines of South Australia* (Adelaide, 1863).

Branagan, D., *T.W. Edgeworth David: A Life* (National Library of Australia, 2005).

Clarke, Rev. W.B., *Remarks on the Sedimentary Foundations of New South Wales* (Sydney: Government Printer, 1878).

Greene, M.T., *Geology in the Nineteenth Century* (Cornell University Press, 1984)

Jukes, J.B., *Sketch of the Physical Structure of Australia* (London: T and W Boone, 1850).

Mawson, D., *Geological Investigations in the Broken Hill Area* (Adelaide, 1912).

Young, R., *This Wonderful Strange Country: Rev. W.B. Clarke, Colonial Scientist* (2015).

Physical Science

Foster, W.C., *Sir Thomas Livingston Mitchell and his World 1792–1855* (Sydney: Institution of Surveyors New South Wales, 1985).

Ingleton, G.C., *Charting a Continent* (Sydney: Angus and Robertson, 1944).

Kass, T., *Sails to Satellites: The Surveyors General of NSW (1786–2007)* (Department of Lands, 2008).

Lines, J.D., *Australian Paper: The Story of Australian Mapping* (Fortune Publishing, 1992).

Mackay, D., "The Mackay Aerial Survey Expedition, Central Australia, May–June 1930", *Geographic Journal* 84 (1934), p. 511.

Astronomy

Explorers of the Southern Sky: A History of Australian Astronomy, eds. R. Haynes, R. Haynes, D. Malin and R. McGee (Cambridge University Press, 1996).

McCready, L., Pawsey J., and Payne-Scott, R., "Solar Radiation at Radio Frequencies and its Relation to Sunspots", *Nature* 158 (1946), p. 339.

Sullivan, W.T., "Early Years of Australian Radioastronomy", in *Australian Science in the Making*, ed. R.W. Home (Cambridge University Press, 1988).

Physics

Home, R.W., "History of Science in Australia", *The History of Science Society* 78 (1982), p. 336.

Home, R.W., "The Problem of Intellectual Isolation in Scientific Life: W.H. Bragg and the Australian Scientific Community 1886–1909", *Historical Records of Australian Science* 6 (1984), p. 19.

Home, R.W., and Needham, P., *Physics in Australia to 1945* (University of Melbourne, 1990).

Roughley, T.C., *The Aeronautical Work of Lawrence Hargrave* (Sydney: Government Printer, 1937).

Engineering

Anything is Possible: 100 Australian Engineering Leaders (2019).

Sharp, S., "Lessons from History: The Contribution of John Whitton", *The Australian Railway Historical Bulletin* (1999), p. 19.

"William Henderson 1896–1978", *Biographical Memoirs of Fellows of the Royal Society* 25 (1979), p. 318.

Wonders Never Cease: 100 Australian Engineering Achievements (2019).

The End Game

Longair, M., *Maxwell's Enduring Legacy: A Scientific History of the Cavendish Laboratory* (Cambridge University Press, 2016).

"Reflecting Diversity: Fellowship and Women in the Australian Academy of Science", *Historical Records of Australian Science* (CSIRO Publishing).

Index

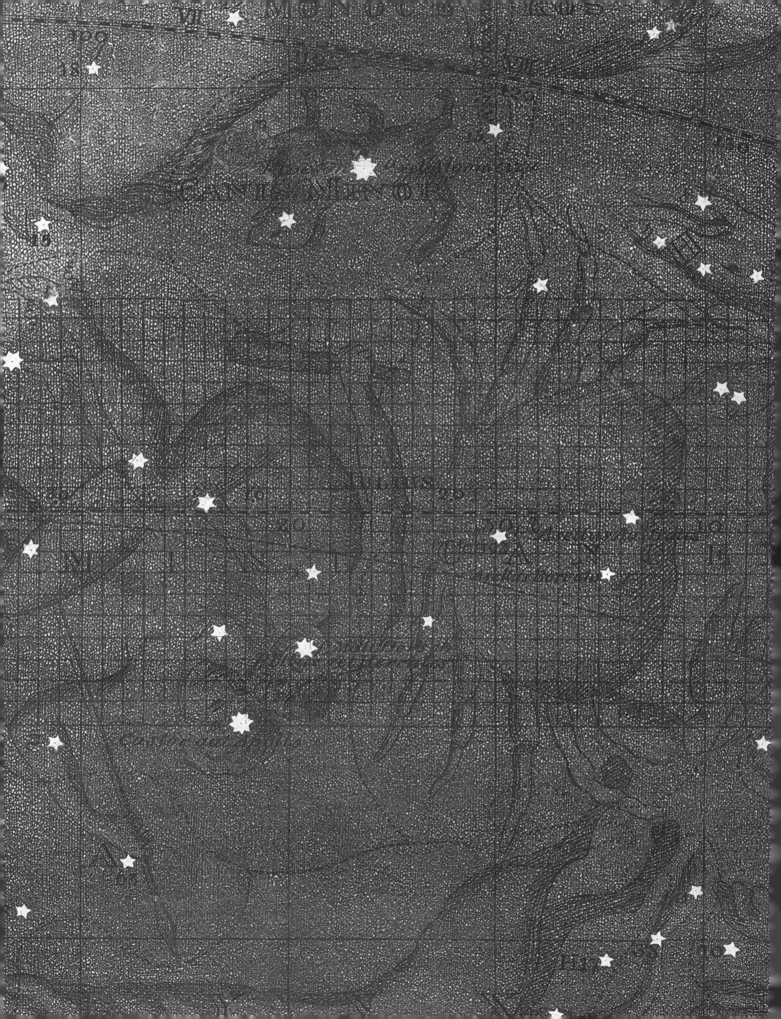